IMAGINING THE FUTURE

Yuval Levin

Imagining the Future

Science and American Democracy

New Atlantis Books

Encounter Books · new york · london

First edition published in 2008 by Encounter Books,
an activity of Encounter for Culture and Education, Inc.,
a nonprofit, tax exempt corporation.
Encounter Books website address: www.encounterbooks.com

Manufactured in the United States and printed on acid-free paper.
The paper used in this publication meets the minimum requirements of
ANSI/NISO Z39.48-1992 (R 1997) (Permanence of Paper).

FIRST EDITION

LIBRARY OF CONGRESS CATALOGING-IN-PUBLICATION DATA

Levin, Yuval.
Imagining the future : science and American democracy / by Yuval Levin.
p.   cm. — (New Atlantis books)
Includes bibliographical references and index.
ISBN-13: 978-1-59403-209-7 (hardcover : alk. paper)
ISBN-10: 1-59403-209-2 (hardcover : alk. paper)
1. Science—Political aspects—United States. 2. Science—Forecasting. I. Title.
Q175.52.U5L48 2008
303.48´30973—dc22
2008032355

10   9   8   7   6   5   4   3   2   1

CONTENTS

FOR MY WONDERFUL WIFE, CECELIA

The future is unknowable, but the past should give us hope.

Winston Churchill

Introduction

THIS IS A BOOK about the place of modern science in the political and moral life of our republic. That is a large subject. It is much too large for a large book, which might be mistaken for a futile attempt to cover it all. It is better suited to a small book, with no pretension to be comprehensive, but which merely takes up one portion of the subject from one angle of approach, in the hope that it might point toward the core.

The portion taken up in this small book is the question of America's political debates about science, and what they tell us about America, about political debates, and about science. My contention is that our kind of politics and modern scientific thought are profoundly tied together in a host of meaningful ways, so that democratic debates about science, and particularly about biotechnology, turn out to be uniquely revealing and instructive.

In recent years, science has been the subject of much controversy. From stem cell research to global warming, human cloning, evolution, and beyond, political arguments have raged. It is polite and customary to bemoan these debates, and to wish that science might not become a hostage to politics. But I believe science raises some crucial social, ethical, and political questions for our society; that far from becoming a hostage to politics, science threatens sometimes to overwhelm our institutions of self-government; and that political debates are the most legitimate way to work through these problems. Our political debates about science, however, have tended not to be very well focused on these problems, and in many instances have tended to obscure rather than clarify the deepest questions at stake.

This book is an effort to show how and why that has happened, and what those questions might be. It therefore begins with an attempt to lay out the nature of the challenge that modern science, for all its immense benefits and promise, sometimes poses to our democracy and to our way of life. It then argues that the political debates we have about science tend to be symptoms of the problem science poses, rather than efforts to address it, and that the actual dispute regarding science in our politics lies deeper than these public arguments suggest. The dispute, I further contend, is not between a culture of science and some competing segment of our culture, but rather between two elements of the larger society (which we would not be remiss to call the left and the right), with science merely one subject of the argument. Neither side is quite "pro-science" or "anti-science," but each has a very different way of understanding what science and technology are for and about. The dispute between them, I then argue, comes down most fundamentally to two quite different ways of thinking about the future, and that difference lies at the heart of a great deal of our political life, well beyond the science debates. Several types of scientific developments in recent years, and especially advances in human biotechnology, have brought this difference to the surface in an unusually sharp and clear way, so that in the arguments about biotechnology (and in a series of other disputes about science as well) we have an opportunity to learn some crucial lessons about our politics. I conclude by suggesting what some of these are, and what hidden dangers lurk in the science debates for the left and for the right.

In taking up one sliver of so large a subject, some limiting decisions must be made at the outset, and one in particular is worth a word. In one form or another, I take up most of what might be considered the science debates of the past few years, but I focus especially on those surrounding human biology and biotechnology. I do this for three reasons. First, the bulk of American science today, and a growing majority of American scientists, are focused on questions of biology and biomedicine. Biology has become the queen of the sciences, as chemistry once was, and physics and mathematics before that. Second, biology and biotechnology raise the deepest moral, social, and political questions. The science of life, especially when directed to man, carries with it both the greatest promise and the greatest dangers, and inevitably also brings forward the most vexing and most heated public controversies. Third, biology in its current form is also the science most nearly approximating the aspirations and designs of the fathers of modern science in the seventeenth century. Francis Bacon, René Descartes, and their immediate successors saw their work as directed above all to the preservation of health and of life, and believed this would be the greatest prize their science would deliver. In some respects, science had been on an extended and most fruitful detour until roughly the middle of the twentieth century—a detour necessary to make the age of biotechnology possible. Today, as never before, the dreams of the fathers of science correspond to the aspirations of practicing scientists, and this gives the scientific project as a whole an unprecedented intellectual and philosophical coherence, and makes the founding ideals and early attitudes of science (which we will explore) powerfully relevant again. Modern biology exemplifies modern science. For these reasons, most of our public debates about science are (and seem likely to continue to be) about biology and biotechnology. I do in places take up some of those that are not (like the global warming debate, for example) but I focus the greatest attention on public disputes about the life sciences.

In the end, however, this book is about our country's political and moral life more than it is about science. It employs the science debates as a lens through which to learn about America, and it is driven above all by a love of America, and a continuing, indeed a growing, amazement at her ways. The book therefore avoids, I hope, unreasonable fears and dark apocalyptic specters. It is not a jeremiad. Some of the powers modern science gives us are surely daunting and immense, and in the hands of a reckless or a foolish people would make for a terrible tragic drama—as indeed we have seen in the history of the last century, and may someday see again somewhere. But in America, science like every-thing else is largely domesticated and democratized, allowing us to gain from its gifts, while what remains of its formidable dark side is turned above all into a challenge to the human character. It is a serious chal-lenge, but one we can meet if we remain true to our traditions.

What I worry about in this book, therefore, are ways in which sci-ence might undermine our ability to remain true to those traditions. And what I seek is an understanding of the character and strengths and weaknesses of American self-government, as revealed in its encounter with science. That, in fact, is precisely where we begin.

The Moral Challenge of Modern Science

IN THE WINTER OF 2002, in the course of a long speech about health policy, President George W. Bush spoke of the challenge confronting a society increasingly empowered by science. He put his warning in these words:

> The powers of science are morally neutral—as easily used for bad purposes as good ones. In the excitement of discovery, we must never forget that mankind is defined not by intelligence alone, but by conscience. Even the most noble ends do not justify every means.[1]

In this familiar formulation, the moral challenge posed for us by modern science is that our scientific tools simply give us raw power, and it is up to us to determine the right ways to use that power and to proscribe the wrong ways.

The notion that science is morally neutral is also widely held and advanced by scientists. Indeed, many scientists wear their neutrality as a badge of honor, presenting themselves as disinterested servants of truth who merely supply society with facts and tools. They leave it up to others to decide how to use them. "Science can only ascertain what is, but not what should be," Albert Einstein said, "and outside of its domain value judgments of all kinds remain necessary."[2]

This proposition seems at first perfectly reasonable. The universe, in its benign indifference, is as it is regardless of what we think is right, and it would seem not to pick sides in moral disputes. Science uses knowledge of the natural world to inform us or empower us, but what we do with that knowledge and power remains up to us.

The most common contemporary critiques of science on moral grounds, moreover, are actually critiques of some uses of technology, and so tend to support this view of science as a neutral tool. Our age of technology has taught us to be wary of the dangers of certain applications of science as tools of manipulation, degradation, or destruction. Any Westerner would recognize the image of Dr. Frankenstein's monster gone wild, and we have all become accustomed, as well, to the specter of the nuclear mushroom cloud, the dread of biological or chemical attack, and the stench of industrial pollution. We have learned the hard way that Dædalus, the mechanic, can be a dangerous character. And we also know that otherwise beneficial technologies can open up troubling ethical questions, and that these will only grow more vexing in the coming years as biology becomes increasingly a science of production just like physics and chemistry before it.

This has been clear from the start. It was Francis Bacon, a father of the modern scientific project, who said plainly in 1609 that "the mechanical arts are of ambiguous use, serving as well for hurt as for remedy."[3]

But Bacon answered his (and President Bush's) worry in terms that still suffice as a reply to the notion that technology's moral neutrality makes it dangerous. "If the debasement of the arts and sciences to purposes of wickedness, luxury, and the like, be made a ground of objection," Bacon wrote, "let no one be moved thereby, for the same may be said of all earthly goods; of wit, courage, strength, beauty, wealth, light itself and the rest."[4] Anything can be turned to evil in the hands of evil men. This is not the most essential moral challenge posed for us by modern science.

The moral challenge of modern science reaches well beyond the ambiguity of new technologies because modern science is much more than a source of technology, and scientists are far more than mere investigators and toolmakers. Modern science is a grand human endeavor—indeed, the grandest of the modern age. Its work employs the best and the brightest in every corner of the globe, and its modes of thinking and reasoning have come to dominate the way mankind understands itself and its place.

We must therefore judge modern science not only by its material products, but also, and more so, by its intentions and its influence upon the way humanity has come to think. In both these ways, science is far from morally neutral.

THE IDEALISM OF SCIENCE

The modern scientific project was not conceived or born as a morally neutral quest after facts. On the contrary, launched in the seventeenth century out of frustration with the barren philosophies of the European universities, modern science was a profoundly moral enterprise, aimed at improving the condition of the human race, relieving suffering, enhancing health, and enriching life.

Francis Bacon argued that a search for knowledge driven solely by "a natural curiosity and inquisitive appetite" would be misguided and inadequate, and that the true aim of a genuine science should be "the glory of the Creator and the relief of man's estate."[5] Man is in need of relief, Bacon suggested, because he is oppressed by nature at every turn, and through his science Bacon sought to master nature and thereby to ease suffering and empower humanity to act with greater freedom.

René Descartes, who stands shoulder to shoulder with Bacon among the fathers of science, had an equally moral purpose in mind. His mathematical science, he informs us in the *Discourse on Method*, aims not at neutral knowledge or the creation of frivolous mechanical toys, but principally at "the conservation of health, which is without doubt the primary good and the foundation of all other goods of this life."[6]

This fundamental moral purpose has always driven the scientific project, and especially the very sciences President Bush referred to in his warning: biology and medicine. Such a moral aspiration may be less obvious in the case of some other sciences, but it is no less significant. Modern science generally seeks knowledge for a reason, and it is a moral reason, and on the whole a good one.

Today, science is still driven by the moral purposes put forward by its founders, and often scientists' very protestations of neutrality attest to this. Consider one prominent example. In a much heralded assessment of the scientific and medical aspects of human cloning published several years ago, the National Academy of Sciences claimed to examine only the scientific and medical aspects of the issues involved while, as the report put it, "deferring to others on the fundamental moral, ethical, religious, and societal questions."[7] This is a fairly routine example of the claim to offer only neutral facts, for judgment by others. But the study concludes by recommending that human cloning to produce a live-born child should be banned because it is dangerous and likely to harm the individuals involved. This, the report implied, is not a moral but a factual conclusion.

In truth, however, it is a conclusion that takes for granted the moral imperatives of the scientific project, and does not even think of them as moral assumptions. After all, why does the fact that a procedure is dangerous mean that it should not be practiced? Does the answer to this question not inherently depend upon a moral argument? Why, if not for moral reasons, do we care about the safety of human research subjects or patients? For that matter, why, if not for moral reasons, do we wish to heal the sick and comfort the suffering? We all know why, and the researchers and physicians engaged in the pursuit of knowledge in biology and medicine know why, too. One imagines many of them chose their occupation in large part precisely because they saw in science a way to help others, and they were right.

Science—and again I speak mostly but by no means exclusively of biomedical science—is driven by a profound moral purpose. Certainly many scientists are moved by a powerful desire to know and to discover and explore. But many at the same time also are moved by the desire to do good—to help the human race by providing us with new knowledge that will translate to new power, and especially power over our natural limitations and afflictions. This purpose does not itself emerge from scientific inquiry, but it guides, shapes, and directs the scientific enterprise in every way. By presenting itself as morally neutral, science sells itself far short.

Many of us nonetheless think of science as neutral because it does not match the profile of a moral enterprise as understood in our times. Put simply, science does not express itself in moral declarations. It is neutral in the very way in which neutrality is seen to be a good thing in a free liberal society: science does not tell us what to do. It takes as its guides the needs and desires of human beings, and not assumptions about good and evil. Our desire for health, comfort, and power is indisputable, and science seeks to serve that desire. It is driven by a moral imperative to make certain capacities available to us, but it does not enforce upon us a code of conduct. It can therefore claim to be neutral on the question of how men and women should live.

But a project on the scale of the modern scientific enterprise cannot help but affect the way we reason regarding that fundamental moral question. Modern science, after all, involves first and foremost a way of thinking. It is founded upon a new way of understanding the world, and of bringing it before the human mind in a form the mind can comprehend. This method of understanding must necessarily leave out some elements of the subjects it examines that do not aid the work of the scientific method, and among these are many elements we might consider morally relevant.

In short, modern science forces itself to consider only the quantifiable facts before it, and from those facts it forms a picture of the world that we can use to understand and overcome certain natural obstacles. The more effectively the scientific way of thinking does this, the more successfully and fully it persuades us that this is all there is to do. The power and success of scientific thinking therefore shape our thinking more generally.

Only when we understand modern science as an intellectual force can we begin to grasp its significance for moral and social thought. The scientific worldview exercises a profound and powerful influence on what we understand to be the proper purpose, subject, and method of morals and politics.

THE PRIMARY GOOD

As he wrote the earliest chapters in the story of modern science, Descartes had already grasped the nub of the matter. Determined, as we have seen, that his new science should be directed to the advancement of health, the notoriously doubtful Descartes was awfully bold in describing health as "without doubt the primary good and the foundation of all other goods of this life."[8]

Surely the claim that health is the primary good has consequences well beyond the agenda of the scientist. Any society's understanding of the foundational good necessarily gives shape to its politics, its social institutions, and its sense of moral purpose and direction. How you live has a lot to do with what you strive for.

And health is an unusual candidate for "the primary good." It is surely an essential good—without health, not much else can be enjoyed. But Descartes' formulation, and the worldview of modern science, sees health not only as a foundation but also a principal goal; not only as a beginning but also an end. Relief and preservation—from disease and pain, from misery and necessity—become the defining ends of human action, and therefore of human societies.

This is a modern attitude as much as it is a scientific attitude. In the ancient view, as expounded by Aristotle, political communities were necessary for the fulfillment of man's nature, to seek justice through reason and speech. Man's ultimate purpose was the virtuous life, and politics was a requisite ingredient in the hopeful and lofty pursuit of that end.⁹ But Machiavelli launched the modern period in political thought by aiming lower. Human beings gather together, he argued, because communities and polities are "more advantageous to live in and easier to defend."¹⁰ The goals that motivate most human beings are safety and power, and men and women are best understood not by what they strive for but by what they strive against. His followers agreed. For Thomas Hobbes, relief from the constant threat of death was the primary purpose of politics, and in some sense of life itself. John Locke, a bit less morbid, saw the state as a protector of rights and an arbiter of disputes, with an eye to avoiding violence and protecting life.

This lowering of aims, then, seems to be as much a result of political as of scientific ideas. But it is no coincidence that Hobbes and Locke were not only great philosophers of modern politics but also great enthusiasts of the new science, just as Aristotle was not only the great ancient philosopher but also the preeminent scientific mind of the Greek world.

Aristotle saw in nature a repository of examples of every living thing in the process of becoming what it was meant to be. This teleology —the understanding of things by their ultimate purpose—naturally informed his anthropology and his political thinking as well: he understood mankind by the heights toward which we seemed to be reaching. The moderns, by contrast, saw in nature a brute and merciless oppressor, always burdening the weak and everywhere killing the innocent. This dark view of life inspired them to aim first and foremost for relief from nature's tyranny. In that way freedom, another word for relief, became the aim of politics, while power and health became the goals of the great scientific enterprise. Rejecting teleology in both science and politics, they understood men by thinking about where they came from—the imaginary state of nature, or eventually the historical crucible of evolution—and not where they were headed.

Avoiding the worst, rather than achieving the best, is the great goal of the moderns, even if we have done a very good job of gilding our gloom with all manner of ornament to avoid becoming jaded by a way of life directed most fundamentally to the avoidance of death. We have gilded it, above all, with the language of progress and hope, when in fact no human way of life has ever been more profoundly motivated by fear than our modern science-driven way. Our unique answer to fear, however, is not courage but techne, which is much less demanding. And so our fear does not debilitate us, but rather it moves us to act, and especially to pursue scientific discovery and technological advancement.

This modern attitude runs to excess when it forgets itself—mistaking necessity for nobility and confusing the avoidance of the worst with the pursuit of the best. From the very beginning, the modern worldview has given rise to peculiar utopianisms of various stripes, all grounded in the dream of overcoming nature and living, at last, free of necessity and fear, able to meet every one of our needs and our whims, and able, most especially, to live indefinitely in good health. This brand of utopianism generally begins in a benign libertarianism, though at times it has ended in political extremism, if not in the guillotine.

But in its far more common and far less excessive forms there is much to admire in this peculiar response to the cold hard world, and we have in fact been very well served by this fearful and downward-looking view of nature and man. Avoiding the worst is in many respects a just and compassionate goal, because a society directed most fundamentally to high and noble ideals inevitably leaves countless of its people behind to face precisely the worst that human life has to offer. Modern societies, egalitarian and democratic, aiming first at relief, put up with far less misery than their predecessors and are far better at practicing genuine compassion and sympathy. And modern life, through modern science above all, has put an end to a great deal of pain and suffering and so has made possible a great deal of human happiness.

As we have done so, we have persuaded ourselves that fighting pain and suffering is itself the highest calling of the human race, or at the very least a foremost purpose of society. The moral consequences of this preeminence of health and relief are quite profound, if not always obvious. A society in pursuit of health is not necessarily a society that neglects the other virtues. On the contrary, the hunger for relief from pain tends to encourage charity and sympathy, and to re-inforce the drive to equality, fairness, and fellow-feeling. Modern societies have been uniquely protective of the basic dignity and inalienable rights of individuals, and of human liberty. The pursuit of health does not necessarily encourage higher and more noble pursuits, but it also does not necessarily conflict with them. Thus, modern life, shaped as it is by the outlook of modern science, can generally co-exist with the virtuous life, shaped by older, "pre-scientific" ideas and aspirations.

But in our time, more than any other in the modern period, we have begun to see the darker moral consequences of the preeminence of health. The pursuit of health does not *necessarily* conflict with other virtues and obligations, but in those cases when it does conflict with them it tends to overcome them. And so when the pursuit of health through science and medicine conflicts with even the deepest commitments of modern life—to equality, to rights, to self-government, or to protection of the weak—science and medicine typically carry the day.

This conflict between primary goods plays out in our contemporary debates about biotechnology—whether embryonic stem cell research, genetic screening of embryos, drug experimentation in developing countries, or any number of others. Almost any violation of human dignity or nascent life can be excused if it serves the purpose of advancing medical science or ameliorating physical misery. It is very hard for us to describe something higher than health, or more important than the relief of suffering, so when relief comes at a cost, even the cost of cherished principles or self-evident truths, we are all too often willing to pay the price.

Moreover, if health and power over nature are the highest human goods, then surely science (as opposed to politics) must be the primary instrument of our fulfillment. Science, far more than politics, directs itself squarely to advancing those goods, and to the extent that politicians try to govern science, they may interfere with that great purpose. For this reason there has long been an inclination to see science as beyond the reach of politics—an inclination encouraged by the fathers of modern science, and one that has established itself firmly in our political mindset. This inclination is perhaps the most fundamental threat to self-government in our time, and among the most profound moral challenges posed by the modern scientific project.

SCIENCE AND SELF-GOVERNMENT

There are, of course, different ways for politics to exert authority over science. To distort or hide unwelcome facts—that is, to manipulate the findings of scientific investigation for political ends—is surely a dishonest and illegitimate tactic. But to govern the practice of scientific techniques that threaten to violate important moral boundaries is not only legitimate but in some cases essential. After all, science is not merely observation. A great deal of science is action, and some of that action (especially when human beings are acted upon) may threaten genuine harm. Politics exists to govern action, and so at times it must govern science. This is not always a controversial point. No one contends that protections of human subjects from violations of their rights in scientific research, for instance, are illegitimate. We argue, rather, about when they are appropriate and to what extent. Because such rules normally exist to serve the cause of safety, they are not deemed to be political or moral strictures on science, but of course that is exactly what they are, and their general acceptance proves the point that the governance of science is sometimes legitimate and necessary.

But when proposed limits are rooted in something other than safety or health—that is, something other than the very same cause science itself serves—they quickly become controversial. And even many limits grounded in a broader understanding of the protection of life tend to be roundly rejected if they place genuine obstacles before the advancement of health. The preeminence of health therefore not only shapes the goals of the scientific enterprise, but also limits the ability of politics to act in the service of other important goods. If the question is whether the advance of science or the authority of liberal-democratic self-government is to prevail, a shrewd gambler would be wise to bet on science.

The defense of scientific freedom in these instances generally takes the form of a defense of free inquiry, and the distinction between mere observation and action is too often ignored. Science, as the servant of the highest good, is deemed to be above politics, and described by its defenders as an agent of truth, not of action. Any subject on which science speaks or acts therefore comes to be seen as off-limits for policymakers informed by other kinds of analysis: by moral premises, or tradition, or religious or personal views, as if every question of public policy with any scientific dimension must be understood as a matter of pure science alone.

Eleanor Holmes Norton, the District of Columbia's delegate to the U.S. House of Representatives, gave voice to this view at a hearing about the use of the abortion drug RU-486 in 2006. Observing that the Food and Drug Administration (FDA) had said the drug was safe for women to use, Norton argued that this conclusion should end the debate, and she noted with regret

> the unmitigated politicization of the one area that Americans always held off from politics, and that is science itself. Whether [Terri] Schiavo or creationism, renamed Intelligent Design, or stem cell research or, God help us, global warming itself, there are views floating around this Congress that essentially reach conclusions on these matters of huge scientific moment, based on their own personal beliefs.[11]

Once science has spoken, Norton suggests, there is no longer any room for "personal beliefs" drawing on non-scientific sources like philosophy, history, religion, or morality to guide policy. "Is it safe?" is the only moral question that science alone can attempt to answer; and so long as something is safe, then all other moral concerns, all other grounds for the governance of science, are deemed illegitimate. Scientific judgment, with health as both the primary aim and only conceivable limit, is the final voice of authority. And it is often precisely those with some political authority, like Norton, who take this view. "My own view," Senator Arlen Specter of Pennsylvania said in 2007, "is that science ought to be unfettered."[12] Responding to a presidential veto of a bill to loosen funding limits on embryonic stem cell research, Illinois Senator Barack Obama told reporters, "the promise that stem cells hold does not come from any particular ideology; it is the judgment of science, and we deserve a president who will put that judgment first."[13]

This elevated view of the authority of science as the chief interpreter of truth poses a profound challenge to the basic liberal tenet of self-government. It delegitimizes other sources of wisdom about what is good and what is not, and it causes citizens and leaders to imagine they simply do not possess the requisite expertise for self-government, and that science would do better. This kind of surrender before the power of the technician is not new. In 1962, for instance, John F. Kennedy told a White House audience:

> Most of us are conditioned for many years to have a political viewpoint, Republican or Democratic, liberal, conservative, moderate. The fact of the matter is that most of the problems, or at least many of them, that we now face are technical problems, are administrative problems. They are very sophisticated judgments which do not lend themselves to the great sort of "passionate movements" which have stirred this country so often in the past. Now they deal with questions which are beyond the comprehension of most men.[14]

But the range of questions assumed to be "beyond the comprehension of most men," and so beyond the legitimate reach of a democratic politics has been expanding as the reach of science into the human realm has expanded. This tends to build a distance between science and politics that progressively narrows the circle of self-government, not only because science is seen to be purely a search for truth, and so immune from limitation, but also because the questions it takes up are seen to be beyond the intellectual grasp of the public and its leaders. In both ways, this is a profoundly anti-democratic tendency.

Two great forces have been building their strength since the seventeenth century: scientific knowledge and public opinion. In an ideal world, our scientific knowledge of nature might inform the opinions of the common man, while the values of citizens might govern the reach of science. But it doesn't take a cynic to realize that conflict between these two great forces is inevitable. In principle, self-government allows the people to reserve the right to exercise their judgment as they wish, and so to respond to the latest pronouncements of science with a "so what?" and make decisions based upon those "personal beliefs" that Delegate Holmes Norton so derisively dismisses. Public policy can and should be informed by all manner of influences. But in our time, on a great many questions, none can speak with the authority that science has. Representative Ted Strickland, Democrat from Ohio, spoke for many when, in a congressional debate about human cloning several years ago, he said that when it comes to issues that touch science, "we should not allow theology, philosophy, or politics to interfere with the decision we make."[15] Not even politics can interfere in politics when science is involved.

OUR MORAL FORGETFULNESS

In part, the supposed supremacy of scientific authority is rooted in the fact that science builds its understanding cumulatively—so that it always knows more today than it knew yesterday. This is not, strictly speaking, how religion works, or in most cases even how philosophy works. Science is inherently progressive, and so gives us the sense that all other means of understanding must strive to catch up. Not far behind every new development in biotechnology is a well-meaning hand-wringer mouthing the all-too-familiar cliché that "science is moving so fast ethics just can't keep up."

But this is a profound misunderstanding. The ethical framework we need to deal with the challenges (and to make the most of the promise) of science and technology need not be developed in light of the latest scientific journal article. Its key components have been available to us for a very long time. They were discussed among the priests in the temple of Solomon three thousand years ago, debated in the markets of Athens in the fifth century B.C., preached by a Galilean carpenter to all who would listen, and they have been and continue to be refined, sharpened, and applied by some of the greatest minds of Western civilization ever since. Our problem is not that we are lacking in ethical principles, but rather that we are forgetful of them.

Modern science and technology stand to exacerbate and worsen this forgetfulness, both by freeing us of some of those things that now and then make us remember—the child whose potential is a great surprise to us, the limits that respect for others must place upon our vanity, the truths and lessons we can only learn by growing old—and by accustoming us to a mode of thinking and learning that always seems to know more today than it knew yesterday. Rightly enamored of the possibilities and achievements of forward-looking science, we are often blinded to the possibility of progress through remembrance, and tempted to believe that we can rise beyond the limits and constraints that the past always seeks to remind us are necessary. This forgetfulness risks leaving us knowing much less than we knew yesterday, even about science.

Science, after all, is a human activity, even if it is one that addresses itself to the natural world. And our civilization has a deep and ready well of knowledge about how to understand and govern human activity. The surest way to understand the role of science in our society, and the surest ground for governing science when necessary, is to resort to that knowledge of human nature and human affairs which is not itself scientific.

Scientific knowledge must of course inform our understanding of the human significance of what science is up to—we need to know what, for instance, the human embryo is in scientific terms in order to know how to regard it in human terms. But science does not resolve the question. It informs the decision, but it is that other great modern force, public opinion, itself informed by a wide array of wisdoms, that sets society's course. Some public defenders of science understand this, devoting great energy and vast resources to winning over the public to their view of the good, as we have seen in the public campaign for embryonic stem cell research. That view of the good, to put it simply, is that uninhibited scientific freedom and generous public funding for scientific research will give the public what it wants most: material progress and cures for nature's afflictions.

But the public, for all its hunger for cures, has other hungers as well. And while the public reveres science, it cannot look to science alone for guidance about what to desire and how to live. It must rely as well upon an assortment of non-scientific wisdom that is perhaps best understood as *tradition*. Tradition, too, is cumulative, but not in the simple sense in which scientific knowledge builds on itself. Tradition is the result of countless centuries of trial and error in human affairs, but it is deeply shaped by the simple fact that human beings always begin in the same place—born helpless, ignorant, and innocent—and always must be shaped and reared to rise from there. Tradition therefore cannot hope simply to build upon itself, because it must shape every generation from the same crude beginnings, regardless of how well its parents were shaped. Our institutions of tradition —cultural, civic, religious, and moral—are therefore always engaged in the Sisyphean task of education, and so are always in some sense doing the very same thing they have always done. They learn from those who have done the same thing in the past, and they seek to improve on those precedents, but they are never free just to move on and do something different, unless they abandon the task of bringing children into the world, and thus abandon the future in the name of progress—a paradoxical shortsighted futurism that cannot last very long.

Our key social institutions are in this sense inherently conservative, and so they must remain. Stability and continuity, which hardly matter much in scientific knowledge, are essential to the cultural vitality of any society.

That does not mean we do not learn new things about how we should live—that our tradition does not evolve and grow. It surely does, and always should, but it cannot do so in a simple and cumulative way. The new things we learn in philosophy and ethics and religion do not supersede the old things we have long known. Modern astronomy has simply proven that what Aristotle theorized about the nature of the solar system was wrong. Modern philosophy will never be able to show any such thing with regard to what Aristotle theorized about the best way to live.

All of this is just to say that there is more than one legitimate way to gain understanding. Our means of understanding and governing man and society are fundamentally quite different from our means of understanding and mastering nature. To understand nature takes ever-growing knowledge. To understand man takes the wisdom of the ages. That wisdom, as it builds, can be informed by scientific knowledge, but it can never be replaced by it. Science is a tremendously effective and powerful means of gaining knowledge about nature, and knowledge of nature is very important. But human beings and human societies are more than mere objects of nature, and so science alone cannot suffice.

Science, morals, religion, and philosophy are not merely different ways of answering the same questions, to be compared to each other based upon their answers. They are, rather, different ways to answer different questions. Modern science, in answering critical questions about the natural world, has brought us health, comfort, wealth, and power undreamt of in earlier ages. These great gains have understandably caused us to concentrate on the sorts of questions science can answer, and so in some measure to lose sight of those it cannot. In this sense, the moral challenge of modern science is a consequence of the power of science to define the questions we ask and the means we seek for answering them, sometimes flattening or deforming what we do and how we live.

IF WE CAN PUT A MAN ON THE MOON...

By its very success and its impressive power, the scientific mindset convinces us that it is the path to the only knowledge worth knowing. We are quite rightly impressed by the effectiveness of scientific methods when applied to nature, and so the impulse to apply the same ways of thinking to non-scientific questions is nearly irresistible. If we can make such remarkable progress in our mastery of nature through science, why should we not make similar progress in our mastery of social, political, and moral questions through science? Science just seems to offer a more advanced way to reason than the old approaches to all of our difficulties.

It has seemed so for a long time. More than two centuries ago, the boosters of science already argued quite openly that scientific thinking would, and should, crowd out other ways of thinking. In his 1794 *Sketch for a Historical Picture of the Progress of the Human Mind*, the Marquis de Condorcet noted that "the sole foundation for belief in the natural sciences is this idea, that the general laws directing the phenomena of the universe, known or unknown, are necessary and constant. Why should this principle be any less true for the development of the intellectual and moral faculties of man than for the other operations of nature?"[16]

Half a century later, Auguste Comte, the father of modern sociology, argued that "the general situation of the sciences of politics and morals today is exactly analogous to that of astrology in relation to astronomy, of alchemy in relation to chemistry, and the cure-all in relation to medicine."[17] With time, he hoped, morals would advance along the path toward more scientific methods. By offering the example of a new and spectacularly effective way of thinking, modern science would replace other ways of thinking, including traditional moral reasoning.

Early critics of science also noted this possibility, though of course with less enthusiasm. Jean-Jacques Rousseau, in his 1750 *Discourse on the Arts and Sciences*, argued that the sheer multitude of objects to which the new learning turned its attention would tend to squeeze completely out of consideration those matters (like morals) to which it did not apply itself.[18] Michel de Montaigne had seen this coming long before, and made the case with his usual terse flair: "the more that men only labor to stuff the memory, the more they leave the conscience and the understanding unfurnished and void."[19] A factual understanding of mechanism, albeit crucial, could not be the only sort of truth worth knowing.

These thinkers, writing early in the modern age, understood the power of ideas to mold our minds, and they could see that the power and effectiveness of scientific methods could dominate the thoughts of men to the exclusion of other ways of thinking.

In our time, we are perhaps less inclined to recognize science as a set of ideas with aspirations to universality precisely because the scientific enterprise has been so successful. But the authority we cede to science, both as the servant of health and as the master of knowledge, weakens our allegiance to those other sources of wisdom so crucial to our self-understanding and self-government. Those other sources serve to ground our moral judgment, while science avoids or flattens moral questions, since it cannot answer them and rarely needs to ask them. Rather than as morally neutral, then, we might describe the modern ascendancy of the scientific worldview as morally neutralizing, crowding out our means of moral reasoning and sources of moral authority. For all its power, science risks leaving us morally impotent.

SCIENCE AND ETHICS

This puts George W. Bush's warning in a rather different light. As the ability of science to remake the natural world continues to expand, science itself, or at least our concession to its authority, has left us increasingly powerless to decide how best to use our novel mastery. The problem is not that our inventions might be used for both good and evil purposes, but that we denizens of the scientific age are at risk of becoming unable to distinguish between good and evil purposes.

Moral imperatives, including especially those profound moral imperatives at the root of the scientific enterprise, are becoming clouded over just as the scientific enterprise begins to focus its attention most directly on the human animal itself.

This leaves science less capable of deciding how it should apply its power, and it leaves society less capable of properly governing the scientific project. Science from the outset has sought not only to know but also to do. The question is: To do what? Without resort to informed moral judgment, the answer, which used to be "to do good," slowly comes to be "to do what can be done." In this way the means of science come to be confused with its ends, the progress of research becomes an end in itself, and we move from the imperative to seek the power to do what we know is good to the notion that whatever we have the power to do is good. "We have bricks, so let us build a tower," [20] we say to one another in the scientific age.

This has never been a very good argument for building a tower, but it has always been a hard one to resist. As science becomes able not only to reach into the skies but also to reach into the human genome and the sources of life itself, we are in greater need than ever of the very moral powers that the success of science has made weaker.

All of this, however, does not mean that science is immoral. Quite the contrary is the case, and this is vital to remember. The problem we confront is so vexing and difficult precisely because science can do so much good, and wishes and aims and attempts at every turn to do so much good. Our challenge is to keep science true to its original moral purpose, while not letting its approach to the world make us blind to moral meaning and judgment. To do this, we must come to understand science as a moral endeavor, a human project with discernible ethical purposes. Only if we see science in such terms, and if scientists themselves do too, can we begin the difficult task of assessing the moral goods at stake, and asking if the good that science can do is in every instance "the primary good and the foundation of all other goods of this life," as Descartes so confidently asserts that it is. There will be many cases when it is deemed to be just that, but there will also be cases when scientific freedom and even scientific progress must be superseded by higher moral goods.

This, in the deepest sense, is the challenge presented to our society by modern science: to advance the great moral good of relieving man's estate while remaining ever mindful of other, and perhaps greater, moral goods. It is a challenge to our sense of what matters most, to our commitments to equality and self-government, to our appreciation of the necessarily varied sources of wisdom and authority, and to our grasp of the right questions to ask.

The real challenge lies not in the tools that science gives us, but in the attitudes it forms in us. The trouble is not that technology can be used for both good and evil, but that people in the age of technology may have real trouble telling the difference between the two. The moral challenge of modern science is, like every genuine moral challenge, a hazard to the souls of men; and the danger that confronts us in the scientific age arises not from our tools or our machines but from our own assumptions and attitudes. To benefit from the advance of modern science while averting the serious difficulties it can carry with it will require an impressive feat of self-government, and yet precisely that advance tends to cut us off from the sources of instruction we will need for such self-governance. This, put most simply, is the difficulty science poses for our politics.

Our political debates about science are replete with symptoms of this difficulty. And the most prominent such symptom may be the very tenor and tone in which these debates tend to be argued—a tone of urgency and crisis, demanding to be treated apart from the conventional preoccupations of politics and government. Having begun to see something of the character of the challenge (and the promise) of modern science for our kind of politics, we are perhaps in a position to understand the source of that peculiar and distorting tone of argument, and so to see a little deeper into the science debates.

CHAPTER TWO

The Crisis of Everyday Life

In April 2007, the United States Senate debated a bill to permit federal funding for research using newly derived human embryonic stem cell lines—and so in essence to encourage the destruction of embryos for research. The outlines of the ethical dispute were quite familiar. Opponents said the embryos, which they took to be developing human beings, should not be treated as raw materials for research. Supporters countered that the embryos could not be considered human persons, and therefore that concerns about their fate should not stand in the way of promising research. But by far the most prominent line of argument was not about the ethical issues at the heart of the debate, but about the potential utility of the embryos in question, and the horrors they might help to overcome. The debate, more than anything, was a discussion of human vulnerability to nature's wrath, and of the terrible injustice of disease. *Washington Post* reporter Dana Milbank described the debate this way:

> "I sustained an episode with Hodgkin's lymphoma cancer two years ago," disclosed Arlen Specter (R.-Pa.). "As a child I suffered from polio," confided Mitch McConnell (R.-Ky.). "My

wife is a breast cancer survivor, my brother died of a stroke, my sister died of an aortic aneurysm," offered Chuck Grassley (R.-Iowa). "I just lost my uncle in Huntington, West Virginia, last year to a form of cancer," submitted Tom Carper (D.-Del.), currently on crutches himself.

On and on they went. Debbie Stabenow (D.-Mich.) invoked her grandmother (Parkinson's), Sherrod Brown (D.-Ohio) cited unidentified kin and his best friend (diabetes), Frank Lautenberg (D.-N.J.) mentioned his grandson (asthma), Patty Murray (D.-Wash.) recalled her father (multiple sclerosis), and Alzheimer's was linked to both New Jersey Democrat Bob Menendez's mother and Tennessee Republican Bob Corker's father.

Then came Gordon Smith (R.-Ore.) with a medical trump card. "I watched my grandmother," he said, "die of Parkinson's. I watched my uncle, Addison Udall, die of Parkinson's. I watched my cousin, former Democratic presidential candidate and Arizona congressman Morris K. Udall, die of Parkinson's."

With yesterday's debate on the Senate floor, the Oprahfication of American politics is nearly complete. While it is difficult to imagine Daniel Webster rising to discuss his cirrhosis of the liver, or Henry Clay requesting floor time to expound on a relative's gout, the 110th Congress has turned the highly personal into the intensely political. . . .

Those too private to discuss their kin mentioned their constituents. Tom Harkin (D.-Iowa), sponsor of the embryonic bill, repeatedly displayed a poster of a twelve-year-old diabetic posing with an array of syringes. Ted Kennedy (D.-Mass.) displayed posters of another juvenile diabetic and a paralyzed veteran. Sam Brownback (R.-Kan.) hung posters of beneficiaries of adult stem cell treatments, including one Parkinson's sufferer who went on safari and "scrambled up a tree to avoid being run over by a rhino." Barack Obama (D.-Ill.), too busy to make a poster or give a speech, tried to get into the act by issuing a written statement about a two-year-old with cerebral palsy.[1]

It is not surprising, or even unreasonable, that the stem cell debate should be carried out this way. The debate is ostensibly about the use and destruction of living human embryos in research. But it gained its force and took on great prominence because it raises the prospect of medical progress and, as we have seen, few prospects speak more powerfully to our imagination, with good reason.

But there is certainly something quite striking about the underlying tone of these appeals to stories of suffering and sickness. It is not quite the case, as some have argued, that these are efforts to overwhelm argument with raw emotion, or to score political points on the backs of the sick and the vulnerable. Some of that surely happens. But if you listen closely to the kinds of appeals described in the passage above, what you find is an anguished plea for justice. Indeed, the pang of anger we feel when we find that a loved one has been stricken by a grave disease may be less a product of grief or sadness than of a wounded sense of justice—a sense that the person we cherish has been treated less well than he deserves. Why this person, at this time? It is the kind of question that might once have had an answer in the form of anthropomorphic gods of place—whose familiar sorts of jealousies and squabbles might at least have offered, in very general terms, something of a recognizable façade of accessible reasons for the cold inhuman ways of nature. But in our time, when we can only look to an inscrutable God or, perhaps even more likely, to a science of cold chance occurrence, the question hits us with all the more power: Why her? Why him? Why now? The question demands an answer in human terms, in terms of fairness and just cause. And when we receive no answer, we feel the need to act somehow, to address the injustice.

This tone of activism is readily apparent in the atmospherics of our various public battles against disease. In our walks against cancer and runs against heart disease and marathons for diabetes research, patients and loved ones literally march in defiance against various ailments, to show that they are stronger than the illness and to raise funds to combat it. These are the powerful theatrics of an American fight for justice, modeled on the efforts of assorted social movements, and especially the movement for racial equality.

But the demand for justice from nature, which is always problematic, is especially so in our times, precisely because when we speak of justice we most often mean equality, and equality is a standard that nature is uniquely unsuited to meet. By some more aristocratic standards, nature can be said to be just—treating the great well and the low poorly—and indeed nature itself can almost be a standard for justice. But if all are to be treated equally, then certainly nature is unjust in the extreme, since it treats people unequally for no apparent reason.

If nature is unjust, then nature must be fought and made to treat us properly. Modern science from the beginning has taken up this cause, and has understood itself to be fighting a desperate battle against a cold and ruthless killer of innocents. In such a fight—a struggle for our very lives—all stops are pulled out, and all tactics are permitted. The fight against disease is an emergency; it requires urgent attention and total focus.

This is the tone of research advocates in public debates about science. They argue that time is running out, but that swift action can still save the day. Testifying before a Senate subcommittee in 2003, Parkinson's patient and research advocate James Cordy told the Senators:

Please, please don't let time run out for me and the over 1.5 million Americans with Parkinson's, and the over 100 million Americans with diseases and conditions who are almost certain to benefit from regenerative medicine, including embryonic stem cell research. It is unconscionable to let time run out—especially now that the scientists tell us the finish line might be within sight.[2]

We cannot let up even for a moment, not for any reason, and especially not now. This is the essence of the argument for approaching medical science with a sense of urgency and crisis: right now is the moment that counts, and we must not let anything distract us. The "finish line" is near.

But if the fight against disease writ large, indeed the fight against natural death, is an emergency, and if at the same time it is a struggle we can never expect fully to win—for surely only the most wild-eyed of our utopians can imagine that Cordy's finish line will ever truly be crossed by medical research, and sickness and death be left behind for good—then we must always live in a state of emergency. We should be always in crisis mode, always pulling out all stops, always suspending the rules for the sake of a critical goal. And that means, in effect, that there should be no stops and no rules, only crisis management and triage. This is the logic underlying a great portion of the arguments heard in our science debates.

Under crisis conditions, we allow ourselves to do things we would never otherwise contemplate. In triage mode, we ruthlessly select among the living to help those who have the best chance at survival. For the sake of saving life, even the most observant Jew can violate the Sabbath. But if life is always at risk and we are always in crisis, then we must always do things that moral contemplation would suggest are wrong. If we are always in a mode of triage, then we must always choose the strong over the weak because they have a better chance at benefiting from our help. And if we must always be engaged in saving life, then we are always justified in breaking the Sabbath, so that in effect there is no Sabbath, no time for rest and contemplation of the truth. Indeed, there is no everyday life at all, against which times of urgency might be measured. There is only the struggle, only the crisis.

Every objection to any means we employ can be answered on the grounds that the struggle demands it, and the struggle is paramount. To stand in the way of the struggle in this moment of crisis, for whatever reason, is to become complicit in nature's iniquity. "Those in a position of advice or authority who participate in the banning or enforced delays of biomedical research that could lead to the saving of lives and the amelioration of suffering are directly and morally responsible for the lives made worse or lost due to the ban," Stanford biologist Irving Weissman told a Senate Committee in 2004.[3] In the heat of crisis, after all, you are either part of the solution or part of the problem.

A WORLD ON FIRE

This logic and rhetoric of crisis lies beneath much of modern life, but it is prominent especially in our debates about the limits of biomedical science, where the stakes of the struggle are most evident. How we got here takes us back again to the origins of our modern way of thinking about nature, and about politics as well.

To the ancients, the normal and the everyday were the measure of things. Man was that creature that could speak and contemplate and seek after truth, and his greatest need was for a means of doing so. Nature was an ordered whole that offered examples of order and wholeness. Science was for contemplation; politics was for finding ways to live well. This approach had its advantages, but it put up with an awful lot of injustice, natural as well as man-made.

The modern approach began with a determination not to put up with such injustice, and so it took on politics and science very differently. To us, the extremes, not the norm, are the measures of things. Man is the creature that can be wounded or killed and therefore needs protection. Nature is best understood when it is stretched and pried and tested under stress in the laboratory or the thought-experiment (in Francis Bacon's paradoxical formulation: "the nature of things betrays itself more readily under the vexations of art than in its natural freedom").[4] Normal life is only a particular manifestation of forces and laws that are best understood when stressed to the extreme, so the extreme, not the normal, is the standard of judgment. This approach, which we associate most with the modern scientific method, is in fact everywhere in our thinking, and has given shape to our understanding of politics and ethics no less than of experimental physics or chemistry. In the science debates, it shows up not only in the ways in which we prioritize ends, but also in how we reason about means.

Consider one very common example, drawn once again from the embryo research debates of recent years. In an effort to challenge the views of those who argue that human embryos are human beings in their earliest stages of development and should not be destroyed for research, Harvard political scientist Michael Sandel raises the following hypothetical (which, as he acknowledges, has been widely used by others in the debate as well):

> Suppose a fire broke out in a fertility clinic, and you had time to save either a five-year-old girl or a tray of twenty embryos. Would it be wrong to save the girl? I have yet to encounter a proponent of the equal-moral-status view who is willing to say that he or she would rescue a tray of embryos.[5]

This, Sandel contends, means such proponents don't really believe that human embryos are human beings. But it is hard to see just how the logic of this argument could justify the treatment of human embryos as raw materials to be destroyed for research. Say you were in a room with your spouse and a complete stranger, and a fire broke out. If you were only able to save one of them, surely you would rescue your spouse and not the stranger, and no one could blame you. But would that then give you the right to go around killing strangers on purpose to take their organs for research? Is that not the logic of the fertility clinic hypothetical? The problem, in other words, is with applying the logic of a building on fire—the logic of triage and emergency—to everyday life. Our world is not a burning building or a lifeboat. To believe that it is would be to deny the legitimacy of almost every ethical and moral limit on action, if that action were directed to addressing the emergency. And if our human nature or our mortal condition is the emergency, then almost any action—any means—would be morally permissible.

"The typical solution of modern[s]," Harvey Mansfield has written, "is to assimilate the normal situation to the extreme, and thus, following Machiavelli, to make necessity the guide for all situations."[6] Taken too far, this easily becomes a recipe for a politics of constant crisis, which is a politics devoid of ethics.

AN ETHIC OF ACTIVISM

This way of thinking not only justifies means, it also has a lot to do with how we think about the ends we seek. Modern science is oriented to action in a way its ancient predecessors never were, precisely because it conceives of the world it studies as the scene of an emergency. And modern politics follows suit. Science is for protecting us from nature; politics is for protecting us from each other.

In principle, both modern science and modern politics are devoted to avoiding pain and averting death. These are decent aspirations, to be sure, but not very high ones. They leave little room for more elevated aims, and, as we have seen, they also stack the deck when the quest to avert death comes into conflict with higher ethical or moral aspirations drawn from other sources. And they also seem to flow from an assumption that fending off the worst is the best that we can hope for. Progress can be never-ending, in this sense, because it never finally succeeds, so the emergency is never overcome. "The felicity of this life," wrote Thomas Hobbes in 1651,

> consisteth not in the repose of a mind satisfied. For there is no such *finis ultimus*, utmost aim, nor *summum bonum*, greatest good, as is spoken of in the books of the old moral philosophers.... Felicity is a continual progress of the desire, from one object to another; the attaining of the former being still but the way to the latter.[7]

The modern ideal, in this sense, has more to do with fending off evil than enjoying good—it is a "pursuit of happiness," not happiness itself, and the pursuit is a form of combat. There is, as Hobbes puts it, no repose.

Combined with our deep devotion to equality, moreover, this all adds up to a recipe for constant urgency and an unending struggle to set things straight. The battle against inequality also sends us to the edges of life, and leaves us struggling to adjust the middle to accommodate the ends, so that no one is left out. Taking our bearings from the extreme case, our society always strikes us as insufficiently accepting of differences, while our heightened sensitivity to inequality means that as the actual remaining injustices grow less and less significant, our outrage against them grows more and more acute. "The desire for equality always becomes more insatiable as equality is greater,"[8] Tocqueville noted. And so it is with health as well. With every victory, the struggles that remain seem all the more pressing and urgent.

The sense of injustice we feel at the sight of a gravely ill child or the inexplicable loss of a loved one is both profound and understandable. It is also nothing new; it is at least as old as Job. But our response to it, the call to national mobilization, the marshalling of troops and arms, the sense of urgency and crisis, the demand to put aside all qualms at least until the battle has been won, these are relatively new. And in this arena, too, every victory makes the next fight seem more, not less, imperative and critical. There is never a lull after success, never a quiet afternoon, never a peace dividend. There is no everyday life in light of which we might define our morality. There is only the provisional morality of crisis: people are dying, this is no time for moralizing.

But the tragic fact is, of course, that people are always dying, and that they always have been and always will be. If this means that there can never be a time for moralizing, then we are in trouble. And the tenor of our debates over the limits of science does suggest that, to many, that is indeed what the facts of disease and of death are taken to mean. Because the whole of the human experience remains imperfect, the whole is taken to be broken, and only the effort to heal it is taken to be worth our time.

A MODERATE MODERNITY

The tone of the science debates therefore tells us quite a lot. It is a function of both the subject matter involved and the underlying attitudes that give shape to both our science and our politics. For anyone who worries that science may sometimes overwhelm our institutions of self-government, this combination is a serious problem indeed. But the nature of the problem also suggests the nature of a solution. The answer is not a rejection of the scientific way of thinking that has brought us so much material comfort and progress; and neither is it a rejection of the way of modern politics, which has given us more freedom and equality, and with them probably more human happiness, than any of the schemes of the ancient political thinkers, and any of the institutions of the ancient polities. Turning back is not an answer, because the progress we have made since the seventeenth century is a prize to treasure and build upon, not an error to reject.

But to build upon it we will need to secure the strength of its roots, and that strength draws on older sources of wisdom that should not be obscured or undone by our progress. The modern project is in some respects a victim of its own success, coming under attack by its own progeny, and some of its most appealing features—the commitment to equality, the celebration of self-government—may be at risk. To be made secure, it needs to be normalized, to be made more (though it can never be made fully) compatible with a way of life that does not measure itself by the furthest extremes and does not live in permanent crisis but that seeks some stability and continuity. It is, in other words, in dire need of moderation.

Most critics of our modern excesses have tended to think moderation an inadequate answer, and to argue that we now confront a choice between decadence and martyrdom: that the only answer to the excesses of the modern enterprise is to flee from it, and to avoid the dark downside of progress by avoiding the benefits too. But this is itself an excessive crisis-mode kind of reasoning, and it miscasts the choice we confront.

After all, there has also been a strong and influential strain of thinking that has argued we can benefit from the advance of modern knowledge as long as we never forget the unsavory realities of human nature, and the constraints placed upon us by the human condition. The great hope of these most sensible modern thinkers—men like Edmund Burke, Adam Smith, Alexander Hamilton, and James Madison—is that we can welcome modern progress without utterly losing ourselves in the process. We could do this by seeing (in this case) science as another human occupation, not the messianic be-all and end-all of human existence. These hard-headed men were not entirely right, to be sure, but they have offered us a way to live well in modern times that for all its failings so far has not simply failed us. They understood the dangers of the modern mindset in excess, but also the value of the kind of progress modern science and politics promise.

Their solution was to insist that our modern projects not put out of mind entirely our ancient ideals and traditions, and that the importance of continuity in politics—of a connection between a people's past and present—be reflected in the practices of our forward-looking societies. And they hoped, too, that room would be made for this approach to coexist with a modern and progress-seeking attitude. This could only work, however, if the tone of urgency and crisis is moderated, and we seek means to advance the ends of the modern way of life in ways that do not run roughshod over our traditions.

Here again, we might turn for an example to the most heated of our recent science debates—the argument over embryo research. In one respect, given the ethical issues in question and the promise of the research, the stem cell debate seemed to force our country into an impossible choice between the sick and the defenseless. The choice was crucially important, and yet it was almost impossible to choose well. Neither our suffering neighbor nor the microscopic nascent life in the lab could be abandoned without grave consequences. Both sides of the argument, moreover, could plausibly present themselves as struggling for the defense of life, and thus both could appeal to our contemporary vitalism, with its adamant parlance of crisis. But the solution, which even now is still emerging, is to see that we need not necessarily accept the tragic all-or-nothing formulation of the proposition: we need not see the killing of the embryo as the only way to extend the life of the patient. Taking our situation to the extreme, in other words, may not be the best way for us to understand the choice we confront, or to choose well. In the everyday world, the world not always in the grip of some terrible crisis, there may be other ways. We will not even seek those other ways, however, unless we see that the task we are engaged in—the task of contending with death and disease—is a permanent aspect of the human condition, not a momentary crisis to be overcome. Neither the patient nor the embryo can live forever, so that the moral question we confront must have as much to do with how to live as with how to avoid dying. Only by grasping that this is the permanent norm can we incline to seek medical progress in ways that also respect moral ideals beyond safety and health.

Most opponents of embryo research, after all, have not suggested that we abandon the quest for cures and relief. They seek rather to pursue it in a moral way. And since the quest will never end and immortality will always be beyond our reach, it is good that we are not confronted with the stark alternatives of abandoning the quest for moral reasons or abandoning morality so we may pursue better health. Instead, we face the moral and political (and technical) challenge of keeping our pursuit of health contained within ethical limits. It is perhaps a more prosaic challenge, but it is an essential one for us today, and the contemporary embryo debates show us why that is.

Fortunately, in the context of the embryo debates, it seems there are also scientific avenues for such a moderate solution, like techniques of producing cells with the abilities of those derived from embryos but without the need to use or to harm human embryos. That researchers have found them is very fortunate. That our society has sought them, however, is the more important indication, for it shows that we have not fully succumbed to the logic of crisis and triage, and that moderation has a chance, if we understand the need for it.

But the fact that the technical solution was so readily available in this case made it easy to avert the terrible choice. The early signs, before alternative stem cell technologies became apparent, were not so positive. The way our public debate about embryo research proceeded for much of the past decade, as evidenced by the congressional debate described above, does not give us much hope that future debates—in which, perhaps, a technical solution will not be so readily or quickly forthcoming—will make room for moderation, or will see beyond the logic of the crisis of everyday life.

To help us see beyond that logic will require a deeper understanding of the character of our disputes about science, and of the issues that underlie our assorted public debates about it. The crisis mentality is one symptom of a larger dilemma, and understanding it will help us see further. That larger dilemma is our eventual destination. But before we can describe the difference of opinion that lies beneath our science debates, we must contend with one common misunderstanding of these debates that threatens to obstruct our view. For several decades now, the controversial place of science in the politics of Western democracies has been explained by many observers as a function of the unique culture of the scientific community itself, and of the friction between that culture and an older humanist culture of the academy, struggling with science for command of the world of ideas. This "two cultures" thesis is where we turn next.

CHAPTER THREE

Two Cultures?

IN 2005 AND 2006, I spent much of my time as a mediator between scientists and politicians. I was a member of the White House domestic policy staff under President George W. Bush, and my portfolio included a range of health, science, and biotechnology issues. This required me often to translate scientific questions into the language of practical decision-making, and on many occasions also to explain the basics of American government to scientists seeking to influence policy or to get government backing for their work. It was a sobering experience, and confirmed a pattern visible in our larger debates about science and politics. People of great intelligence entrusted with enormous authority in politics and government lose all their confidence when confronted with a scientific question. On issue after issue, tongue-tied politicians have struggled to make sense of complex scientific terms denoting even more complex scientific concepts, and more importantly have tried to discern what role, if any, politics and the larger culture should have in overseeing science. And serious scientists, with decades of experience, cannot imagine why the public or its leaders would seek in any way to limit or to govern their work.

Many researchers and advocates for science have been genuinely frustrated and puzzled by all the fuss, unable quite to see what concerns are being expressed, and why they matter.

All of this tends to corroborate an old cliché: that a deep chasm separates scientists and non-scientists in the intellectual culture of the West, and that it is this difference of style, of background, of basic assumptions and terms, that explains the character of our public debates about science. Those debates, in other words, express differences between science and the lay society.

In one form or another, this cliché has been with us since the earliest days of modern science; and in its most recent incarnation we have known it as the problem of "the two cultures." That evocative phrase was introduced by British novelist C. P. Snow as the title of his 1959 Rede lecture at Cambridge University. The lecture, later published in *Encounter* magazine, and ever since widely available in book form, quickly gained worldwide acclaim as the definitive description of a profound and important problem, though it also sparked a heated controversy in intellectual circles in Britain and America. Ever since, discussions of the differences between the culture of science and the larger culture have generally begun by summarizing Snow's famous thesis, though the summary may well have come to obscure Snow's actual case.

As commonly retold, Snow's argument is roughly that scientists and humanists read different books, start from different premises, have different habits of mind, seek different ends, and almost speak different languages, to the point that the two camps simply cannot understand each other; and that this is a serious problem. This is a useful synopsis of the old cliché, but it is not a very good description of Snow's *Two Cultures* argument. In fact, Snow's thesis had much more to do with politics, and rested upon premises that are much harder to swallow than the simple story of a cultural divide.

Given the great prominence of the "two cultures" analysis, it is well worth our while to carefully consider Snow's actual argument before we proceed: to review his case, to listen to his critics, and to think about whether his description of the problem helps or hinders our understanding of the contemporary science debates.

SNOW'S TWO CULTURES

Charles Percy Snow was, by all appearances, perfectly suited to point the attention of the world to the gap between the scientific and the humanistic cultures of his time. He had been trained as a chemist at Cambridge, and had done serious laboratory work for several years, before joining the civil service during the Second World War. After the war, he never returned to the lab, but he worked for over a decade in the British science bureaucracy, deciding how public funds should be allocated, and which researchers deserved support. In these years, he also began to publish a series of novels that met with popular success and critical acclaim. By the late 1950s, he was able to leave his government job and make a full-time living as a writer. And by the time of his *Two Cultures* lecture, he was a prominent figure on the British intellectual scene. He was, in his own terms, a member of both cultures.

What he saw from this unusual vantage point worried him deeply. His lecture begins by sketching the outlines of two distinct and separate factions, both within and outside the academy. The scientists, as Snow describes them, constitute a real culture. "Its members need not, and of course often do not, always completely understand each other," Snow writes, "but there are common attitudes, common standards and patterns of behavior, common approaches and assumptions that go surprisingly wide and deep."[1] Among the humanists and literary intellectuals, "the spread of attitudes is wider," and there is less coherence, but they are nonetheless united by "a total incomprehension of science that radiates its influence" throughout this varied group.[2] Snow's description of this other culture focuses on the writers and literary critics whom he knew best, though he clearly has in mind a wider class of humanists, or even just non-scientists. The two cultures, as he sets them out, are therefore the cohesive culture of science, and the broader, more diffuse culture of non-scientific intellectuals.

Snow then offers telling anecdotes of a profound disjunction—a "gulf of incomprehension"—between the basic attitudes and assumptions of the scientist and the intellectual.[3] Most of the scientists he knows have not read Shakespeare, or Dickens, and do not really see why they should. In a gathering of literary types, Snow asks if anyone could tell him about the second law of thermodynamics, or could define mass, or acceleration. Nearly no one could answer. Members of each group lacked an understanding of the very basic building blocks of the other group's view of the world, and so the two simply could not communicate. Each also has a distorted view of the other. "The non-scientists," Snow writes, "have a rooted impression that the scientists are shallowly optimistic, unaware of man's condition." Meanwhile, the scientists view the intellectuals as "totally lacking in foresight" and closed-minded.[4] Both cultures are distorted by their insularity.

This first segment of *The Two Cultures* is generally all we take away from the lecture, and we rightly credit Snow with describing a genuine problem. But the theme of Snow's lecture is not the existence of this cultural divide, but rather its significance and its practical consequences for international politics. In pointing to these, moreover, Snow only seems concerned about one half of the cultural divide he has described.

Having justly rebuked his scientist friends for their lack of interest in the riches of Western culture, and rightfully scolded his fellow writers and critics for their ignorance of basic scientific concepts, Snow makes it clear in the rest of his lecture that only the latter of these two elements of the problem seriously worries him, and should seriously worry the rest of us. *The Two Cultures* is really a lecture about the scientific illiteracy of the literary class. "Intellectuals, in particular literary intellectuals," Snow argues, "are natural Luddites."[5] These intellectuals, whom Snow dubs members of "the traditional culture," in his view never fully understood the meaning of the industrial revolution, and they responded to it with either silence or contempt. By so doing, Snow suggests, they cut themselves off from the future, because the future belongs to the sort of applied science that made industrialization possible.

The scientists, Snow writes, "have the future in their bones," while "the traditional culture responds by wishing the future did not exist."[6] And this dispute has real consequences, because the West has real competition for control of that future. As Snow's lecture enters its final section, it dawns on the reader that the target of his worries is not so much the gulf between scientists and humanists as the gulf between the rich and poor nations of the world, and the battle between the Soviet bloc and the West for influence and power over Third World countries seeking to improve their circumstances.

"One truth is straightforward," Snow writes: "industrialization is the only hope of the poor."[7] And the only route to industrialization is through the straightforward technical application of scientific knowledge to material problems. What the world's poor nations want are, in Snow's words, "men who will muck in as colleagues, who will pass on what they know, do an honest technical job, and get out." Fortunately, he observes, "this is an attitude which comes easily to scientists . . . that is why scientists would do us good all over Asia and Africa."[8] And that, in turn, is why the educational systems of the West must be better geared toward producing scientists. Snow worried that the Soviets had a clear edge in this regard, and they would own the future if the British and the Americans did not seriously reform their educational priorities.

The greatest barrier to such reform was precisely the fact that the "traditional culture" of non-scientific intellectuals lacked a basic understanding of what science is or does, and in some ways was even openly hostile to the scientific enterprise. "It is the traditional culture, to an extent remarkably little diminished by the emergence of the scientific one, which manages the Western world," Snow writes.[9] And yet it is the scientific culture that holds the key to the West's ability to compete for sway over the future. The problem of the gulf between the two cultures, therefore, essentially reduces to the problem of the scientific illiteracy of the governing classes—a problem to be addressed by a reform of Western education with the aim of producing more scientists and fewer humanists.

This peculiar argument, then, is not so much about scientists and humanists, but about the rich, the poor, and the Cold War. In fact, in an essay written four years after his famous lecture, Snow acknowledged that "before I wrote the lecture, I thought of calling it 'The Rich and the Poor,' and I rather wish that I hadn't changed my mind."[10] But if he hadn't changed his mind, the lecture certainly would not have lasted as it has, and would not have drawn the reaction it did.

THE GREAT CONTROVERSY

Indeed, the response to Snow's lecture is an indispensable part of the story of *The Two Cultures*. From the beginning, most commentators ignored the geopolitical theme of the lecture, and focused on the cultural divide suggested by its title. Snow had clearly hit a nerve by articulating the tension between the culture of science and the "traditional culture" in the West. Early published commentary on the lecture generally commended Snow for bringing the problem to light, and on the whole seemed also to share his view that the more pressing of the difficulties that result from the cultural divide is the scientific illiteracy of non-scientists.

But not everyone was so kind. The most famous, or infamous, of Snow's critics was the prominent Cambridge literary scholar F. R. Leavis. His critique of *The Two Cultures*, itself delivered as a lecture at Cambridge in 1962, and later published in the *Spectator*, was scathing to the point of brutality. Indeed, its venomous bite, more than its argument, was what drew most people's attention at first. Leavis said Snow was "as intellectually undistinguished as it was possible to be."[11] He attacked Snow's literary work ("he can't be said to know what a novel is") and said the *Two Cultures* lecture did not suggest that Snow had any particular aptitude for scientific rigor either. Leavis's comments dripped with bitter acrimony, and went well beyond the usual bounds of academic disputes. It was clear that Leavis not only rejected Snow's thesis, but also reviled everything Snow stood for.

But beneath the invective, Leavis lodged an important substantive charge as well. He argued that Snow had grossly misrepresented the character of the literary culture, particularly in suggesting that the English writers of the nineteenth and twentieth centuries had no sense of what the industrial revolution meant. On the contrary, Leavis argued, while these writers may have found much to dislike about industrialization, they understood what it meant quite clearly and deeply. The effort to contend with the industrial revolution was, in fact, a defining theme of English literature for a century, and it was simply foolish to accuse the writers of the time of ignoring the phenomenon. Leavis believed this gross distortion of the literary culture was a symptom of Snow's blindness to the deepest human questions, a blindness he suggested was well apparent in the substance of Snow's argument, and was all too characteristic of the culture of the time. In an age of great and rapid change, we more than ever need to be reminded of our "full humanity," Leavis argued, but Snow's lecture called on the West to abandon it as irrelevant and focus instead on crude material questions alone.[12]

Leavis's belligerent tone distracted most readers from the substance of his case, and the reaction to his lecture was mostly shock at its outrageous breach of manners. The lecture became something of a scandal, and "the Leavis-Snow controversy" remained a burning issue in British intellectual circles for years.

The definitive overview of the controversy, though, came from an American. In an article entitled "Science, Literature and Culture: A Comment on the Leavis-Snow Controversy" published in the June 1962 issue of *Commentary* magazine, Lionel Trilling, a leading writer and critic of the day, offered his assessment of the style and substance of the dispute.

Trilling conceded that F. R. Leavis had gone beyond the bounds of good taste in his assault on Snow. Leavis's tirade had been delivered in "an impermissible tone," according to Trilling, which was both "bad in a personal sense, because it is cruel" and "bad intellectually because by its use Dr. Leavis has diverted attention, his own included, from the matter he sought to illuminate."[13] But on the substance of that matter, Trilling largely agreed with Leavis.

Snow's *Two Cultures* lecture, according to Trilling, was far from an even-handed assessment of the faults of scientists and literary types. In fact, he argues, it was "nothing less than an indictment of literature on social and moral grounds. It represents literature as constituting a danger to the national well-being, and most especially when it is overtly a criticism of life."[14] The writers' critique of the industrial age, which Snow had cast as their grand failure, was, in Trilling's view, among their most valuable accomplishments.

More importantly, Trilling shrewdly diagnosed the deepest faults in Snow's line of reasoning. Snow, he argued, was careless in describing the gulf that separates science from the rest of the culture because that gulf was not exactly what concerned him. "The real message of *The Two Cultures,*" Trilling argued, "is that an understanding between the West and the Soviet Union could be achieved by the culture of scientists, which reaches over factitious national and ideological differences."[15] This attitude, of course, downplayed everything meaningful about the East-West divide, and ignored the most basic and crucial realities of the moment.

Trilling saw that this was the principal weakness of Snow's argument, and that it was very likely a fatal weakness. "It can be said of *The Two Cultures,*" he writes, "that it communicates the strongest possible wish that we should forget about politics."[16] Decades later, that is still what stands out most about Snow's lecture, and today the error is, if anything, more starkly revealing.

SCIENCE AND POLITICS

Nothing is more telling of this anti-political character of Snow's approach than a short footnote in the published text of his lecture.

Following his claim that the scientific culture has the future in its bones while the culture of the humanists wishes "the future did not exist," Snow suggests to his reader: "Compare George Orwell's *1984*, which is the strongest possible wish that the future should not exist, with J. D. Bernal's *World Without War*." [17] This little aside does not reflect well on Snow's judgment, though it does help explain his attitude toward politics.

Bernal's book, published a year before Snow's lecture, is a classic in the genre of painfully naïve disarmament tracts, filled with dreams of material progress achieved by socialist economics, scientific advances, and a utopian politics of peace. Like Bernal, Snow seemed to believe that science offered an escape from the frustrating realm of political principle and argument. The defining challenge of the moment, in Snow's view, was the poverty of the non-industrialized world. International politics would therefore have to address itself to the concrete task of industrialization. Scientists were evidently best-equipped to do this—not only because they had the technical skills, but also because they were most likely to wish to forget about politics and just do the job—and so the future belonged to whomever had more and better scientists. It seemed clear to Snow that if the West did not rise to the challenge, the East surely would, and would dominate.

Our vantage point today allows us to conclude that on this crucial point Snow was mistaken. To a limited extent, both East and West did follow Snow's preferred approach to international development, and abandoned the ill-conceived civilizing mission of European imperialism for more practical policies of technology transfer. The results mostly proved the point that scientific technology is insufficient without humanistic wisdom. In the best of cases, new technology did bring some relief from poverty, though almost never from political despotism. And in the worst of cases, it has brought us barbarians with sophisticated weapons of mass destruction.

More importantly, the West did not lose out to the Soviet Union, even though it was never as committed as the Soviets to industrializing the Third World. Communism collapsed, not for lack of scientists, but because the communist approach to politics and economics, indeed to human nature, proved completely unsustainable. Snow believed the future of a society would be determined by the character and quality of its scientific education. But in fact, the future came to be determined by a society's sensitivity to human freedom and dignity, and to individual rights—that is, by the character of its politics.

Snow accused the "traditional culture" of overlooking the significance of science, and in part he was right. But he was guilty of precisely the equal and opposite oversight. He failed to appreciate the significance of politics, not only for the general future of the West, but also for the particular future of science. The two cultures are in fact deeply dependent on one another.

Indeed, looking back at the Cold War, it is clear that the West fared better not only in economics but also in science. The United States certainly came to place great emphasis on science education in the wake of the launching of Sputnik (and so, even before Snow's lecture) and other disconcerting displays of Soviet scientific competence, but the radical reforms in education that Snow implied were necessary never did come. The Soviets were surely more committed to science, and devoted more resources to producing scientists, than Britain or America. And yet, the West always had the edge in both practical applications and theoretical advances. This should not surprise us. Free societies offer incentives for innovation, and room for different human types to pursue different paths and techniques. Despotisms often make all-too-effective uses of technology, but they rarely make serious conceptual or technical advances. The proper sort of scientific education matters deeply, to be sure, but it turns out that political liberty and free markets are also essential prerequisites for making the most of applied science. In order to truly thrive, science needs liberal democracy.

And liberal democracy needs science too. The West's facility for consistently generating prosperity and ever-higher standards of living was undoubtedly an essential element of its ability to prevail over communism. And this extraordinary aptitude for progress is certainly dependent on the bold and innovative spirit of modern science and technology.

Snow understated the degree to which science depends upon the principles and ideals of the traditional culture; and that culture, as he pointed out in his critique of it, seems to have underestimated the degree of its dependence on the progress of science. Each side failed to understand its reliance on the other precisely because it failed to understand the character, aims, and mode of thought of the other. This is the essence of the "two cultures" dilemma. Snow described one half of the problem, and embodied the other.

But this profound interdependence does not mean that political controversies about scientific advances are a result of a chasm of mis-understanding between science and the larger culture. The "two cultures" problem—in its clichéd form, not in the form Snow laid out—does persist, but it is not at the heart of our public disagreements about science, because the role of science in determining our political prospects is not as central as Snow believed it to be.

THE TWO CULTURES TODAY

The basic contours of the cultural divide Snow described have certainly changed since his day. Snow's subject was something of an inter-academic divide between humanist intellectuals and professors of scientific subjects. It was almost a campus squabble, though one with grave consequences. Today, the scientific culture Snow described is still quite recognizable, though his culture of humanists is not what it was.

Natural and physical scientists, today no less than in 1959, form a fairly distinct cadre whose attitudes and views are sufficiently consistent and self-contained to be reasonably, if crudely, grouped under one broad tent. Perhaps the most significant change in that culture since Snow's day has been the greater infiltration of the profit motive into scientific work. For good and bad, much of the work at the cutting edge of scientific discovery today is done with product development and marketing in mind—whether in academic labs hoping to profit from patents or in private corporate labs seeking the next big thing. Among other consequences, this has introduced a degree of secrecy into laboratory work that is otherwise anathema to the scientific culture, and it has also of course affected the priorities of researchers more generally. In most other ways, however, the culture of science today still resembles that which Snow described four decades ago: it seems unified around a faith in its mission of discovering truth and benefiting humanity; proud of the rigor of its methods and the willingness of its members to be proven wrong and to follow the facts; deeply, not shallowly, optimistic; collegial; practical; and supremely self-assured. The culture of science basks in society's appreciative awe, and is showered with financial support, both public and private. Morale is very high.

This culture of science, in fact, is an extraordinary cultural achievement. As William Galston has put it:

> The community of scientists has furnished an enduring ideal for liberal democratic public life: a model of inquiry among equals, oriented toward truth and human betterment rather than toward power, inwardly protected against the inevitability of individual prejudice and bias, insulated from nefarious external interference, carried out in conditions of fullest possible publicity, and ultimately determined only by the force of the better argument.[18]

But the culture of science is also intellectually insular in the extreme. Perhaps because of the hard-earned status and high morale of their profession, too many prominent scientists think too little of the principles, ideals, institutions, and modes of thought of the non-scientific culture. In responding to challenges from that culture, the defenders of science often seem to be caught up in ancient myths—of Galileo and Darwin, of Truth versus The Church. Again and again, those arguing for science in the public square seem to operate on the assumption that their opponents are advocating ignorance, and are driven by purely religious motives. They seem to have almost no sense at all of what their critics actually argue. Their appeals almost echo some eighteenth-century French *philosophe*: utterly confident of the right-eousness of scientific progress, disdainful toward all critics and traditions, railing against mostly imaginary priests and kings.[19] With some prominent exceptions—who are either genuine political radicals or cautious self-aware moderates—researchers tend not to see why some of their work troubles some non-scientists. Snow saw much of this in his colleagues.

But the culture of humanist intellectuals, and particularly literary intellectuals, has changed dramatically since Snow's observations, and generally not for the better. When Snow delivered his lecture, he could speak of a lively active milieu of writers, critics, and humanist academics who were deeply engaged in the life of the larger culture, and who might even be said to direct that culture. He accused them of despising the culture of science, but it may be more accurate to say, as Lionel Trilling did, that they offered a profound critique of industrialization and the age of technology.

To put the matter bluntly: no such culture exists today. The literary culture in the West—that is, the academic culture of literary theory and criticism—has for some time now been deep in the throes of a profoundly self-destructive fit of foolishness that has made it almost irrelevant to the present and future life of the larger society. Literature matters, as always, but the full-fledged literary culture, with its mix of chic radical politics and self-involved postmodern posturing, simply does not. And it certainly offers no meaningful critique of the age of technology. We have no T. S. Eliot, or even Lionel Trilling, today, who might stand as a true social critic with something to say about the culture of science.

The last substantial confrontation between the literary and the scientific cultures involved the so-called "science wars" of the 1980s and early '90s, in which a segment of the humanistic culture sought to argue that science was purely a social construction, with no greater claim on objective reality than any other such construction. The scientists reacted to the challenge with a mix of bemusement and anger, and the whole thing came to a point when physicist Alan Sokal submitted an article parodying a critique of science to the cultural-studies journal *Social Text*. The article consisted of layer upon layer of meaningless jargon pretending to argue that the theory of quantum gravity showed that science would soon be freed from the tyranny of objective reality. It was a hoax, but the journal's editors fell for it, published the article, and, along with the entire field of "science studies," were embarrassed beyond words when Sokal revealed his cruel joke.

The affair demonstrated above all the utter lack of seriousness in the contemporary humanist engagement with science. It proved, if any proof was needed, that the literary culture, if not the larger culture of academic humanists, was lost in the wilderness, and had little to offer in the way of guidance for living well in the age of science and technology.

Given the choice between today's literary culture and today's scientific culture, any reasonably sensible society would choose to be led by the scientists. But of course, that is not the choice we actually face, and there is certainly more to our larger culture than literary intellectuals and academic humanists. There is a citizenry seeking to make the most both of the great benefits of science and of the moral principles of liberal democracy. This larger culture—the democratic culture, which is still in some ways a "traditional culture"—has its own concerns and ideals, and even some of its own intellectuals. It is today more distinct from the humanist culture than it was in Snow's day. Indeed, it is grossly disserved by the bulk of its humanists, who can certainly no longer be said to manage things, but it has also not consented to be governed by its scientists. Our politics do not, after all, amount to a "yes" or "no" question about science. Indeed, that is precisely where the "two cultures" cliché falls short.

THE REAL SCIENCE DEBATE

The tenor of our political culture, the terms and concepts of our political life, are not the purview of one professional clique or another, but are the common language of society. And the distinction between scientists and non-scientists is not a defining difference among citizens.

The idea that either scientists or humanists govern is an element of Snow's larger rejection of genuine politics, which resulted in his inability to perceive the sources of strength that eventually led to the victory of the West over the Soviets, a victory he had thought impossible unless the West were put fully into the service of scientific and material progress. But in a culture that can see the value of a genuine politics of self-government, the line between scientists and laymen does not turn out to be determinative.

Snow's argument that scientists know too little about their cultural inheritance, and that non-scientists know too little about science, does ring true and looks certain to continue ringing true. It is a function of the character of scientific education, on the one hand, and of the technical complexity and sophistication of scientific work on the other. But that does not mean that this division is itself crucial to the state of modern politics, or even that it explains the place of science in our political life and thought.

Science does occasionally function as an interest group, but the science debates are not properly understood as arguments between scientists and non-scientists. They are, rather, arguments between two segments of our larger political culture, and science is just one subject of their argument. At issue is not a matter of different intellectual training, or professional standards, or scientific literacy—and not a disagreement over facts. At issue, rather, is a serious and comprehensive difference of opinion about society's priorities, obligations, and purposes: a difference of worldview.

Properly understood, then, the science debates are one manifestation of the fundamental difference in worldviews that defines our politics. Science is not the only issue on which adherents of these two camps disagree. Far from it. But their disagreement over science is uniquely revealing of the deeper difference of opinion that divides them more generally. By examining their arguments about science and technology, we can learn more about both politics and science in our time. Those arguments, and those worldviews, are therefore where we turn next.

CHAPTER FOUR

Two Visions of the Future

IT IS A STRANGE and striking fact that the participants in America's science debates seem to break down along the basic contours of our larger political division between left and right. There are exceptions, of course, and the left-right divide itself is not always absolutely clear or consistent. But as a general matter, viewed from a healthy distance, the parties to the science debates are not so different from the parties of our politics more broadly. Many sober observers have taken this to mean that our society's understanding of science and its place has been corrupted by political interests, and co-opted into arguments that have nothing to do with science itself. But perhaps the arrangement of forces in our science debates tells us something more significant than that, and might in fact help us dig deeper into the sources of the left-right divide itself.

The language and the terms of our science debates point beyond themselves to these more profound divisions. This is especially so in the case of the most prominent of the science debates: those involving advances in human biotechnology. For at least three decades, but especially since the late 1990s, the future of these biotechnologies has been a hot political issue in this country. Novel prospects for manipulating nascent human life, enhancing physical or mental powers, reshaping the life cycle, or otherwise exercising unprecedented control over our biological selves have increasingly been fodder for public argument. Both the means and ends of these emerging technologies have been controversial. But advocates and critics of particular advances tend to agree about one thing: biotechnology will play a critical role in shaping the future of humanity.

These debates make it plainly apparent that to think about science and technology is to think about the future. It is, unavoidably, to speculate and to predict, to imagine how our lives might be affected by new knowledge, new tools, new methods, and new powers. Most arguments about technology are therefore really arguments about the future. They give voice to different sorts of expectations about progress and change, and to different sorts of intuitions about the character of human life. The particular technology being debated is often secondary to these larger much-disputed themes, and the public debate is shaped by different ways of imagining the future at least as much as by the specific technical potential of a new device or technique. Most political debates, on any subject, are of course in some general but crucial sense about the future, but the question is much closer to the surface in the science debates, and especially in the biotechnology debates.

At issue are not exactly different sets of predictions. At its extremes, each side in the biotechnology debates may indeed have some specific image of the future in mind, whether of a post-human techno-utopia or of some static nostalgic ideal. But for the most part neither side pretends to know exactly what is coming, and both recognize that the future will not yield any one permanent or stable state but a dynamic and constantly evolving experiment in human living—just like the past and the present. Rather than specific competing predictions of the future, at issue in these controversies are different ways of imagining the future in general, and different ways of thinking about some large and basic questions: What is the future? How do we get there? Who lives there? What matters most about it?

Such questions are rarely taken up so explicitly, of course, but behind the arguments of different partisans in the biotechnology debates there clearly lurk a set of rudimentary assumptions about these very subjects. These assumptions tend to coalesce into two broad schools of futurism: one thinks in terms of future *innovations*, and the other thinks in terms of future *generations*. The differences between them explain a lot about our contemporary technology debates. Each is too easily and too often caricatured by the other, but if taken seriously, each also offers a rich and compelling anthropology of progress—a sense of how the future happens in real human terms.

The biotechnology debates offer a uniquely vivid opportunity to examine these competing anthropologies of progress, and to see whether they point us to a reasonable and recognizable understanding of the human experience, and therefore whether they can be relied upon to guide our thinking about the future.

THE ANTHROPOLOGY OF INNOVATION

To imagine the future in terms of innovation means, most fundamentally, to imagine change in terms of new ideas, and to think of life as an array of individual experiments and choices. It is to ask how we might best encourage innovation, how we might allow the best innovations to flourish (and the worst to be rejected), and how new ideas allowed to thrive can alter human life.

This may be the more familiar and—to us liberal, forward-thinking Americans—the more obvious approach to thinking about our future. For better or worse, the future will be shaped by the innovations and advances of the present: by what we develop, what we build, what we learn, what we discover, what we try and test and deem worthwhile. Progress, in this sense, is made possible by improvements in our knowledge and understanding, our abilities, our circumstances, our institutions, our technology, and our control over nature and chance. There is of course always a danger that we may misuse our newfound powers, or even that they might corrupt us; but there are also reasons to believe that we will learn to use them responsibly, to enhance our lives and improve our world.

Not surprisingly, in the debates over biotechnology this innovation-driven view tends to be favored by liberals and libertarians of all parties—those who oppose restrictions on new techniques and technologies. This is not because they share some simple-minded optimism about biotechnology, but because they share a faith in the processes that drive innovation and progress in a free society, and believe that impeding these processes, or even trying to control them in advance, will only make things worse. They do not deny that serious difficulties may arise as the result of new innovations and technologies, but on the whole they argue that these difficulties can be overcome by the very same method that best serves innovation: trial and error governed largely by individual choice.

Indeed, if we think of the future primarily in terms of human innovation, then this dynamic and unmanaged trial and error process turns out to be the all-important filter that determines what tomorrow will bring. After all, it is usually foolish to try to control or even to predict the course of future developments in science and technology, and so any attempt to govern technology with strict rules determined in advance will probably fail to encourage the best and to prevent the worst. Rather, the way to assure that the best practical innovations ultimately triumph is to assure that new ideas are put to the test of real-world use, so that only those that turn out to be good for us are kept. Those individuals most directly affected by some new innovation will be best able to judge its value, and if they find it is harmful or not worthwhile, they will reject it. This understanding of the future implies that the most constructive and sensible policy regarding the new is to place as few constraints as possible in the way of innovation and as few limits as possible on the individual's power to choose.

The combination of innovation and choice, each feeding back into the other in a self-correcting process, will work in a complex, unpredictable, but highly effective way to secure for us a future that works, even if we could not have imagined it. The future, after all, is *our* future, and so we are likely to make choices and to judge the consequences of our choices in ways that look out for our own best interests, and therefore that seek the best sort of future. As Virginia Postrel notes, in laying out her own engaging version of this view, "by shaping our individual lives, choosing among and arranging the things we do control, we form a larger pattern that is under no one's control, yet is complex and orderly."[1]

This anthropology of innovation is founded in a recognition of the intricacy and volatility of human life, and in the sense that both good and bad ideas may emerge from wholly unexpected sources, so that in thinking about the future we must above all be prepared for the unexpected and make room for it. This means not closing off potential avenues of progress simply because we can imagine how they might lead society astray. We can never really know where anything will lead, after all, and it would be unfortunate to lose out on a possible advance only because we could not have imagined it. "Humiliating to human pride as it may be," wrote the great economist Friedrich Hayek, "we must recognize that the advance and even the preservation of civilization are dependent upon a maximum of opportunity for accidents to happen."[2]

This general vision offers an account of the human condition that we can readily recognize. It is the logic behind much of our liberal democracy, our free-market economy, and our culture of individualism, and so has probably been responsible for more liberty, prosperity, and plain human happiness than almost any other set of ideas in the history of the human race. It is closely akin to the modes of thought that underlie the modern ideal of progress, and it also coincides nicely with the worldview of modern science and its devotion to trial-and-error experimentation, to an unimpeded freedom to inquire and explore, and to a forward-looking faith in progress. It is therefore no surprise that those most adamant about this way of imagining the future are also especially adamant about defending science and technology from regulation or restraint in the political system. Modern science and its progeny are agents of this kind of innovation, which is possible only in an environment that nourishes experimental liberty.

This underlying vision of the future does, however, suffer from two particularly noticeable weaknesses, both of which are especially apparent in the biotechnology debates.

THE LURE OF UTOPIA

The first weakness is an inclination to utopianism, with many of its attendant eccentricities and dangers. This may seem like a peculiar charge to lay at the feet of so dynamic a vision of the future. After all, the anthropology of innovation, even if it yields in glowing prophecies of better days to come, is not quite utopian in the conventional sense, because it usually does not envision an ideal, stable, blissful end-state toward which all innovation is tending. Rather, it imagines an open-ended process of progress, by which new ideas and new knowledge are turned into new power and put in the service of the pursuit of happiness.

Still, as the philosopher Hans Jonas suggested in *The Imperative of Responsibility*, this view may be utopian in a deeper sense, and especially in the context of biotechnology, because it accepts at least as an option the possibility of profound and potentially permanent alterations in the human condition—indeed, in the nature of the human being.[3] The potential of genetic selection or manipulation; of mood, memory, or personality control; of radical life-extension, and similar biotechnological possibilities add up to the prospect of taking our own nature in hand and making it an object of manipulation and design. In practice, this entails alterations of those facets of human nature that have always been the permanent backdrop against which all other change has occurred and been measured, and that have always been the solvents of dangerous utopian fantasies.

Utopian experiments were bound to fail, in Winston Churchill's prescient words, because they were "fundamentally opposed to the needs and dictates of the human heart, and of human nature itself."[4]

Western critics of communism often made this a central tenet of their case. But if our nature is in our hands, and our intrinsic inclinations and desires can be managed, then no such limitations would restrain utopian ambitions—especially if they were only exercised at first at the level of the individual.

In some of its more extreme formulations, the short distance between the innovation-driven vision of the future and utopianism is very easy to see. *Converging Technologies for Improving Human Performance*, a report published in 2002 by the National Science Foundation, offers a glimpse of this sense of the future. The report makes a case for human progress through relatively free technological innovation, and then argues that technologies for radically improving and remaking human performance will initiate a process of "changing the societal fabric towards a new structure." If it is not held back by ignorant critics, the report argues, the convergence of nanotechnology, biotechnology, information technology, and cognitive science may spawn "a golden age that [will] be an epochal turning point in human history." Indeed, it continues, "technological convergence could become the framework for human convergence—the twenty-first century could end in world peace, universal prosperity, and evolution to a higher level of compassion and accomplishment."[5]

Assorted "transhumanists" and "extropians" dream of even greater things, including liberation from the bonds of the body and the possibility of endless life. According to the prominent transhumanist writer Max More, "death is an imposition on the human race and no longer acceptable." Therefore, he continues,

> to Extropians and other transhumanists, the technological conquest of aging and death stands out as the most urgent, vital, worthy quest of our time. . . . Certainly, the achievement of posthuman lifespans will require extensive revision of our way of life, our institutions, and our conception of our selves. Yet the effort is worth it. Limitless life offers new vistas, unexplored possibilities, unbounded self-development.[6]

Indeed, the genuine expectation of conquering death has long been a hallmark of the more extreme formulations of the innovationist approach to the future, and of the hopes it tends to place in modern science. As far back as 1793, the English futurist and writer William Godwin looked forward to intellectual advances that could bring about a "total extirpation of the infirmities of our nature," including not only pain and disease, but also melancholy, sloth, aggression, and hate. At the end of it all, he foresaw the prolongation of human life "beyond any limits which we are able to assign."[7] In their approach to imagining the future, some contemporary partisans of unrestricted biotechnology clearly echo Godwin's prophecy of progress.

But these are extremists, and such views are most certainly the rare exception even among libertarian futurists today. At the conceptual level, of course, what is revealed at the extremes of any movement can often teach us something about what is buried in the center. But it can teach us only so much, and the radical voices at the edges should not be taken to speak for the partisans of innovation more generally.

Most friends of innovation are not such outright champions of a post-human age. Their inclination to utopianism far more often consists of an inchoate readiness to contemplate a radical reworking of the human condition as one potential option for the future. This inclination may demonstrate a lack of moderation, and a willingness (if not an eagerness) to see the future unmoored from the past and the present. These are alarming indications, but in themselves they do not mean that the anthropology of innovation is somehow simply fanatical, or even wrong.

THE MISSING LINK

The second flaw in this vision of the future does, however, pose a significant problem. Put simply, those who imagine the future in terms of innovation tend to think of the future as something that will happen to *us*, and so as something to be judged and understood in terms of the interests of the free, rational, individual adult now living. That person is the basic unit of measurement in all of the theories of social life that inform the anthropology of innovation: the freely choosing individual of classic liberal democratic theory; the rational actor of free-market capitalism; the consenting adult of libertarian cultural theories. All of these models and theories serve us well because enough of us do more or less answer that description much of the time.

But the future is populated by other people—people not yet born, who must enter the world and be initiated into the ways of our society, so that they might someday become rational consenting adults themselves. Strangely, what is missing from the view of the future grounded in innovation is the element of time, or at least its human consequent: the passing of generations. What is missing is the child—the actual bearer of the future of humanity—and the peculiar demands, conditions, and possibilities that the presence of children introduces into the life of our society and its future.

In part, children are absent from this vision of the future because the vocabulary of classical liberal and libertarian thinking leaves little room for them. The thought-experiment that is liberalism's creation myth—that famous state of nature from which free and equal men enter together into society and government for the protection of their rights—holds out a timeless ideal. Government is legitimate because free individuals created it by choice and live under its rules in accordance with a kind of contract. But only the founding generation of any society can claim to have done that. The generations that follow did not freely create their regime. They were born into it, literally kicking and screaming. They enter a world formed by laws, arrangements, and institutions that were established by others, but which they have no real choice but to accept. They are also incapable, for about the first two decades of their lives, of fully exercising the rights of citizens. And yet every decision made by their society will directly affect them and those who will follow them. So by the logic of the theory, how can we take into account the needs and rights of future citizens who are not there to consent? How can we keep from treating them unjustly?

Liberal theorists have not been blind to this difficulty, of course; and more importantly, like many things that occupy political philosophers, these concerns are really far more of a problem in theory than in practice. The theorists come up with complicated notions of implicit consent and implied participation, while in actual societies liberalism is suspended in the family, and parents are trusted to look out for the interests of their children.

Nonetheless, it matters that the theory of the liberal society and the anthropology of innovation have serious trouble with children and with future generations. Our theories do shape our ideals and our actions, and affect our sense of what is legitimate and what is desirable.

The most common answer to the liberal difficulty with the child is to treat children as the charge and almost as the property of parents, and so to apply the language of rights to them second-hand. This often makes good sense, but it also has the effect of subsuming the interests of the child within those of the parents, so that in principle our picture of the world can still consist purely of rational adults and their needs and wants. That way, we can continue to imagine the future without considering the distinctive challenges (and the peculiar promise and hope) that result from the presence of children in society.

But the absence of children in this vision of the future results from more than a gap in a theory. Even more important is the very practical way in which children pose a hindrance to any vision of progress. Regardless of how much intellectual and material progress any society may make, every new child entering that society will still enter with essentially the same native intellectual and material equipment as any other child born in any other place at any other time in the history of the human race. Raising such children to the level of their society is, to put it mildly, a distraction from the forward path. And a failure to initiate the next generation of children into the ways of civilization would not only delay or derail innovation, it would put into question the very continuity of that civilization.

The constant intrusion of children into our world reminds us that even as we blaze a trail into the new and unknown, we are always at risk of reverting very far back into humanity's barbarous origins, because we are always confronted with new human beings who have just come from there. We are, in a limited sense, always starting from scratch, and this means that we need more than innovation to secure and to better our future.

The anthropology of innovation would like to avoid or avert this complicated reality. It does so mostly by ignoring it, but at the edges of the party of innovation, we see genuine efforts to ward off the challenge of the child. In the "transhumanist" desire for eternal life is a desire to think of the future as belonging quite literally to us, and not to future generations. It is a desire to start not from scratch, but from individual, rational, freely choosing adults, and to progress only from there.

Indeed, it may be that in its fullness, this innovation-driven vision of the future almost has to exclude children. William Godwin offers a sense of why that should be. In his future, free of "disease, anguish, melancholy [and] resentment," when people might live nearly forever, progress would almost depend upon the absence of children. "The whole will be a people of men, and not of children," Godwin writes of his utopian ideal, "generation will not succeed generation, nor truth have, in a certain degree, to recommence her career every thirty years." [8]

This may be the only way in which the anthropology of innovation could be sufficient in itself as a vision of the future. But the fact that truth has, "in a certain degree, to recommence her career every thirty years," or in other words that children enter the world knowing nothing of it, is a defining feature of the life of every human society. Children do not start where their parents left off. They start where their parents started, and where every human being has started, and society must meet them there, and rear them forward. That we are all born this way has everything to do with how the future happens.

Hannah Arendt, borrowing a term from the demographers, labeled this inescapable fact of life human "natality," the counterpart of human mortality. [9] A vision of the future that takes note of our natality would lead us to imagine in a profoundly different way.

THE ANTHROPOLOGY OF GENERATIONS

To imagine the future in terms of generations means, most fundamentally, to be concerned for continuity. The means of human biological continuity do not offer guarantees of human cultural continuity, because (at least for the time being) the intellectual and cultural progress we might make leaves no real mark on the biology of our descendants. They enter the world as we did, and as all human beings have before us: small, wrinkled, wet, screaming, helpless, and ignorant of just about everything. At this very moment, dozens of people are entering the world in just that condition—about 15,000 worldwide make their entrance every hour—and the future of the human race depends upon them. Contending with this constant onslaught and initiating these newcomers into the ways of our world is the never-ending and momentous challenge that always confronts every society.

At stake are both the achievements of the past and—most especially —the possibilities of the future. Continuity, not stasis or return, is the essential political and social imperative for conservatives, and the facts of natality are a principal reason. The great late-eighteenth century philosopher and statesman Edmund Burke, a father of modern conservatism, described society as "a partnership not only between those who are living, but between those who are living, those who are dead, and those who are to be born."[10] Adherents of the anthropology of innovation in Burke's time explicitly disagreed with this description. "The earth," wrote Thomas Jefferson, "belongs always to the living generation," and "one generation is to another as one independent nation to another."[11] They argued specifically for a severing of links between the generations, in the name of progress and renewal. But conservatives have argued just the opposite: that the facts of natality mean no such severing is possible, and that attempts at such discontinuity will threaten precisely the prospects for progress. If the task of initiation and continuation fails in just one generation, then the chain is broken, the accomplishments of our past are lost and forgotten, and the potential for meaningful progress is forsaken. The barbarism of savage human nature, more than the prospect of a final human victory over natural limitations, is in this sense always just around the corner, always put before us by the newest arrivals.

Indeed, what stands out about the anthropology of generations is not so much a desire to protect children from the dangers of the world—a desire shared by nearly everyone—but rather the related determination to protect the world from the dangerous consequences of failing to instruct the up-and-coming generation.

As Arendt points out in her classic essay on this subject, "The Crisis in Education," the task of education initiates a new child into an old world, and so is responsible for two things: for the child's initiation and for the world's continuation. It is at once responsible for every individual and for the whole society over time. These two missions are not the same. The child must be protected from the world even as he benefits from its advantages and opportunities. And the world must be protected from the child—from the prospect of savagery—even as it benefits from exposure to the freshness, vitality, and hope of the young. The child is protected in the arms of a family that is in turn strengthened and reinforced by a culture friendly to its cause. And the world is protected through the transmission of culture and civilization.

The very term "culture" already hints at this project. From the Latin *colere*, meaning to tend or to care for (to "*cult*ivate"), culture draws upon an agricultural metaphor that points to the need for the appropriate conditions for growth. The work of the culture is the work of cultivating human souls, providing them with nourishment and with protection as they grow. The culture provides the background preconditions without which a society could not contend with the challenge of natality. This is one main reason why conservatives—to whom the anthropology of generations most appeals—care so much about the culture and its mores.

It is also why some vague and seemingly abstract concerns—like human dignity and human nature—matter so much to conservatives engaged in the biotechnology debates. Such ideas cannot help but shape the way the next generation understands its place and its purpose, and some potential innovations in biotechnology cannot help but affect these ideas. This is why it is critical to think about today's innovations with the future in mind, and to consider their implications for future generations who would enter a world that takes these innovations for granted.

Indeed, this sort of thought-experiment is key to much of the approach of those drawn to the anthropology of generations. When thinking about a world profoundly influenced by some new technology or innovation, they do not ask only "what would it be like to live in that world?" They ask also "what would it be like to enter that world, knowing only that world, growing up in that world, being shaped by that world?" They judge each innovation not only by how it might enhance or degrade their own life, but also by how it might improve or diminish the ability of our society to raise and to tend to the next generation, and by its influence on the inheritance we could leave for the future. They therefore sometimes judge innovations very differently than those who think of the future primarily in terms of the interests of the present.

In fact, this generational approach to the future implies that innovation is not as significant as it may sometimes seem, because the most crucial project of every community remains mostly the same over time. Because the challenge of initiation and continuation is absolutely critical to the survival of every society, the most important thing that any society is likely to be doing at any given moment is educating and rearing the next generation. This is the most important thing human beings did in the past, the most important thing we now do in the present, and the most important thing the human race will need to do in the future. It is obviously not the only thing we do, but it is the essential prerequisite to anything else we might want to do, emphatically including innovation and progress.

The necessary tools for this critical ongoing mission—families, communities, institutions, and cultures that encourage transmission, initiation, and continuity—are therefore permanently necessary, and are generally more important than almost anything else we might imagine when we think about the future. These need to be defended and encouraged, because it is very difficult to conceive of a future without them.

Other important projects we engage in, as individuals and as societies, can be judged in terms of their effects on this imperative goal of perpetuation and transmission. This way of thinking often has a powerfully edifying influence: we feel compelled to live well so that we might provide a model of a life well lived for those who follow. But even when it cannot claim this benefit, this way of thinking keeps us alert to the genuine needs of the future. If some approaches to progress undercut the prerequisites for further progress, they must be understood and judged as such.

This might occur when certain potential innovations stand to seriously undermine our ability to pass along to future generations the ideals, the virtues, the knowledge, the traditions, the living spirit of our society—that is, when innovation stands to alter something so profound about the human experience that the inheritance of the future would be significantly diminished as a result of its loss. These are the sorts of dangers that conservatives in the biotechnology debates are eager to repel.

This eagerness and this worldview, however, are open to two very serious drawbacks, which conservatives are not always sufficiently ready to admit or resist. The first is an exaggeration of the threats to childhood and to future generations, and an excessively protective stance that threatens to turn politics into a branch of pediatrics. The impulse to protect children from exposure to the larger world threatens to suffocate them (and us) if it is not tied to an effort also to initiate and expose them to that world. It is easy to go overboard in childproofing our culture, and it is easy to underestimate the ability of children to contend with and to process cultural influences. Some threats to transmission and to childhood are very real—and some biotechnologies, which reach children at a primal biological level, may pose such threats—but we should not go too far in estimating the vulnerability of the next generation.

The second drawback is a tendency to confuse the project of transmission with that of preservation. This is the conservative version of the utopian impulse. It begins from a tendency to idealize the past, and falls into a self-caricaturing blind nostalgia, and into simpleminded "when I was a kid" modes of argument. These can be found at the edges of the party of generations, just as the post-humanists lurk at the edges of the party of innovation. These conservative extremists are no less misguided than their libertarian counterparts, and no less guilty of missing the point.

The lesson of the anthropology of generations is not so much that the past should be preserved, or even that change should somehow be governed in its every detail. That is not only impossible but thoroughly undesirable. Rather, the point is to recognize that a set of several very basic things—centered especially on the rearing and education of the young—must be allowed to happen in the future, precisely because they are the prerequisites for progress. These can be aided and improved by many human innovations, and left mostly untouched by others. But they might also be significantly undermined or made impossible by certain sorts of innovations, and these must be avoided when they can be. Trial and error alone cannot always be trusted to discern the difference, because the costs of error are too great.

But how, then, can we discern the difference? How do we tell genuinely dangerous prospects apart from merely startling novelties? The costs of erring too far on the side of caution can be very high, especially when innovations in medicine may be at stake. What does the anthropology of generations suggest that we should truly be concerned about in the fast-approaching age of biotechnology? Some examples will gesture toward an answer.

———

CLOSING OFF THE FUTURE

Perhaps the most significant consequence of human biotechnology for the project of transmission and perpetuation is the potential, for the first time in human history, to directly manipulate the raw material of the next generation: to alter and control the biology of our descendants in advance. As the scientific journal *Nature* noted in an editorial following the cloning of Dolly the sheep: "The growing power of molecular genetics confronts us with future prospects of being able to change the nature of our species." [12]

The most fundamental fact of human natality has always been that human nature emerges from the womb in essentially the same general form in every generation; or, as conservatives like to put it, that human nature has no history. The implications of this insight can hardly be overstated. It sits at the core of the conservative understanding of human life and society. It is the reason that those social and political arrangements that have passed the test of time are worth preserving— because the "test of time" is really just a nearly constant repetition (in changing circumstances) of the challenge of promoting human virtues and satisfying human wants in the face of some permanent facts about human beings. It is the reason that new ideas too must be tested against the hard realities of human nature, and, for this reason, it is also the principal solvent of utopian fantasy and totalitarian ambition. The Marxist dream of a "new man" free of the old attachments and desires ran head-on into the permanence of the old man's nature, and was forced to succumb like so many wicked fantasies before it. Human aims and innovations have always had to comport themselves with human nature, and this has generally worked as an effective moderator of otherwise reckless projects.

But what if human nature could instead be made to comport with human aims and innovations? What if rather than reshaping the world to suit man's nature, technology was turned to reshaping that nature —to reshaping man himself? The reeducation camps of twentieth-century totalitarianisms were ineffective (not to mention horrendously inhumane) attempts to do just that. Could biotechnology offer a more effective and more compassionate means? The answer is maybe, and it depends.

It seems unlikely that biotechnology will ever simply allow us to control or to program the psyche of the unborn. But through a combination of some foreseeable advances in genetics, neuroscience, embryo research, and assisted reproduction, along with techniques of screening, selection, and crude manipulation, we could at least come to select our descendants based upon a probability of their possessing characteristics (including some of personality and mind) we find desirable. Technologies developed to screen out disease very easily become available to screen out other traits, and the capacity for manipulation and engineering will likely grow more plausible with time. As we learn more about the underlying causes of aggression, or melancholy, or cognitive ability, or even artistic or musical skill, among countless other traits, we will be better able to screen for these traits in both the genotype and the early phenotype of embryos, fetuses, and children, and perhaps someday be able to design and engineer them in as well.

This new power would carry with it some grave consequences and some heavy burdens of responsibility. We would be responsible for the character of the next generation (and perhaps all future generations) in a way we never could have been before, and at the same time, by plying our influence at the level of biology rather than moral education, we might grossly restrict the liberty of our descendants.

It is very likely true, as the innovationists would remind us, that parents would only choose what they understand to be best for their children. Parents always have. But this point misses the nature and scale of this new technological power. Our sense of what is good and bad for our children is built upon a moral vision of human life that was grounded in the old ways: in response to human nature, and in the expectation of the permanence of that nature. And our ability to act on that sense has always been restrained by the stubbornness of the traits children somehow already possess. In a world of positive control, both of these constraints would be profoundly altered. The edifying humbling limits on the parent's power—and with them the very newness of the child—would be diminished.

That newness would diminish because the next generation, and those that come after, would be less and less surprising to us, and more and more a product of our plans and purposes. As Hannah Arendt put it, in the context of education:

> Our hope always hangs on the new which every generation brings; but precisely because we can base our hope only on this, we destroy everything if we so try to control the new that we, the old, can dictate how it will look. Exactly for the sake of what is new and revolutionary in every child, education must be conservative; it must preserve this newness and introduce it as a new thing into an old world.[13]

Rather than new people in an old world, the generations designed by our biotechnology would increasingly be familiar people—made to suit our preferences—in a new and unfamiliar sort of world, a world defined by a profound discontinuity, unhinged from the limits that defined the past, and so unlikely to bring forth the surprises that define the future: a world living always in this present. The innovationist ideal becomes a self-fulfilling prophecy.

We would also find ourselves stuck with the consequences of present ideas and fads, imprinted permanently in the biology of our descendants. In almost every age, someone has proposed some clever and terrible scheme for how children should be reared and raised. The West's first great philosopher suggested that children should be separated from mothers and fathers, and raised in common by what amounts to a bureau of parenting, and the world has since seen no shortage of similarly bright ideas.[14] Misguided educational fads have done real damage now and then, but they have generally not gone very far, because some traditional practices grounded in natural attachments seem to accord best with the character of parents and children.

Such practices have resisted every effort at radical reform. But direct interventions in children's bodies and minds, and particularly genetic interventions or selections that extend to the germline, would make permanent the preferences of the present, and would subject future generations to our whims. It has been very good for us that the raw material of humanity remains raw in every generation.

By imagining the future in terms of generations we can also see how the imposition of parental preferences, even advantageous ones, could constrain a child's sense of personal liberty and potential. Imagining the future child rationally analyzing neutral facts, *Reason* magazine's Ronald Bailey has written that "the designer babies of the future will have more knowledge and therefore will have a far greater scope for free choice than we do today."[15] But freedom is not just another word for nothing left to know. One's sense of independence would certainly be hampered by the knowledge that one's intellectual faculties or biological features were made to order or chosen off the shelf. Think of what it would be like to enter the world as a person with physical or mental traits selected in advance, and to grow and get to know oneself as such a person. Donor-conceived children today already confront serious challenges of identity and origin; think of what it would mean to know that your parents chose you or exactingly designed you to possess certain qualities, to affect certain traits, to be some particular way.

Not only the knowledge of *which* traits you were chosen to have, but even simply the knowledge that you are as you are because your parents expected something in particular out of you, would be certain to constrain your sense of possibility and independence. It is far from clear if such a child would indeed "have more knowledge" about his or her humanity, or would feel a greater sense of freedom than the countless generations who have spent their lives discovering and revealing their potential.

In purely biological terms, the trait-selected child would still have an unknown potential, because we are not likely to develop anything approaching absolute control of the biology of our descendants. But in terms of the human experience of life, that child, unlike any of us, would live always shadowed by the presence of parental will expressed in his or her own biology. The issue is not some genetic determinism, but rather the concern that the knowledge of having been designed by another for a particular purpose—of being, in a fundamental material sense, what someone else decided he or she should be—would diminish a child's sense of freedom and possibility. We know what can happen when children are pushed too hard to live up to parental expectations and wishes. If that push exerted itself in the child's very biology, its effect (even if only implicit and emotional) would be despotic in the extreme.

And what of that child's own children and grandchildren? This diminution of freedom would intensify as its effects reverberated through the generations. C. S. Lewis understood this consequence of our increasing power over man in 1944, even if he did not foresee the precise technological means of achieving it. In *The Abolition of Man*, Lewis speaks of

> the picture which is sometimes painted of a progressive emancipation from tradition and a progressive control of natural processes resulting in a continual increase of human power. In reality, of course, if any one age really attains, by eugenics and scientific education, the power to make its descendants what it pleases, all men who live after it are the patients of that power. They are weaker, not stronger: for though we may have put wonderful machines in their hands we have pre-ordained how they are to use them. . . . The real picture is that of one dominant age—let us suppose the hundredth century A.D.—which resists all previous ages most successfully and dominates all subsequent ages most irresistibly, and thus is the real master of the human species.[16]

It is no surprise that the present-centered anthropology of innovation, which seeks to ignore the critical task of transmitting our cultural inheritance to the future, has also taken it upon itself to stop the end-lessly reiterating procession of generations, and to take in hand the biology of our descendants, turning the future into an unlimited extension of the present. If the future must be populated by other people, say the innovationists, let them at least not start from biolog-ical scratch. And yet, by unmooring human nature from its permanent foundations—foundations that have been the sources of our social, cultural, and political institutions—this project would indeed start future generations from scratch in a more profound and decisive way.

This is one way in which biotechnology directed to the human person has the potential to dramatically disrupt the all-important process of transmission, and one reason why those informed by the anthropology of generations worry about it. Engineering human bio-logical change is, in these terms, a very different matter from engi-neering animals and plants to better serve our needs. It changes "us" to better serve us. And once it has done so, we are cut off from the roots of all other movements for change and improvement. The modern age and the scientific revolution have sought, with great success, to better fit the world to man. But by altering man himself, we now seek to better suit mankind to . . . what? Only to the short-term wishes of the present. Imagining the future in terms of generations helps us see how terribly shortsighted such a project is likely to be, and how dis-ruptive of the critical mission of bringing up future generations it is almost certain to be.

HUMAN DIGNITY AND THE CULTURE

The mission of managing the junction of the generations relies, as we have seen, not only on the work of individual parents or teachers, but also on some shared sense of the character and significance of a full and dignified human life, and on a culture that supports and builds that sense. The way we understand ourselves obviously shapes the way we introduce ourselves to the next generation, both the lessons we give and the examples we offer.

In the biotechnology debates, this is why conservatives defend large and often fairly vague ideas of human dignity, human limits, and human excellence. For many conservatives, the argument about biotechnology is an argument about the future of our idea of humanity. That idea shapes human ideals and aspirations, in this generation and in future ones; it is the substance of what we stand to teach the future.

In subtle but absolutely critical ways, the biotechnology revolution is likely to impinge on this self-image of humanity, and in doing so to affect the assumptions and intuitions of future generations entering a world reshaped. By changing the way they regard their humanity, it will affect the way they live it out and pass it on.

Our ability to reorder and transform some prime ingredients of the human experience—our desires, our bodily selves, the relation of our actions and our happiness—requires us to think in a new way about the meaning of our innovations for the future. Changes in the relations between parents and children, between effort and performance, between body and soul, could hardly help but influence humanity's understanding of itself and so our very sense of what a human life entails. The question is whether these changes will diminish or enhance the lives lived under their influence.

We should not pretend to have a simple answer to that question. But here again, it is crucial to see things through the eyes of a new generation entering the world we are constructing, and growing up knowing no other. To grow up in a world where personality and behavior are subject to carefully targeted scientific control, where physical performance and mental acuity are routinely enhanced by drugs, where procreation is a laboratory procedure, where the human animal is primarily understood as a chemical machine to be manipulated by a rational controller, is to develop in a very different place than that which has built up our idea of human life and human aspiration until now. It is to mature, and to build the capacity to reason and intuit, in an unfamiliar universe of concepts, where the basics of human being, acting, and feeling in the world stand profoundly altered. No one can know exactly what these changes will mean. But we also cannot simply expect that a rational, humane, or noble choice will mean the same thing to a person who has grown up in such a place, with such a sense of self, as it now does to us. Diminished concepts of human activity, human relations, and human dignity might influence the present generation only mildly, indeed perhaps only theoretically. But the effects on our ability to introduce ourselves to future generations who would grow up knowing no other way would be far more significant.

This worry is painfully vague and notoriously difficult to translate into the language of liberal democratic politics, but it is no less real for being so. It lies at the bottom of a great deal of the general disquiet regarding the age of biotechnology. Rendering it into recognizable social and political arguments is a key challenge for any future conservative bioethics. The language of human dignity begins to point in this direction, and conservatives in the coming years will need to work to make that language more concrete and to understand its implications.

IMAGINING THE FUTURE

These general reflections do not by any means simply add up to arguments for stopping the progress of biotechnology, and the concerns they raise do not simply outweigh the great promise of many biotechnologies. But they do add up to an argument for thinking about the future in terms of those who will actually live there—in terms of future generations.

Thinking in these terms reminds us of the heavy burden of responsibility we bear, as a generation confronting the biotechnology revolution at its outset. Our new and growing power to affect the future of humanity calls for a new reflection on ethical principles. As Hans Jonas understood, our unprecedented ability to affect the nature and the character of future generations means that *responsibility* must be the center of this new ethical approach, in a way that it has never had to be before. This responsibility demands that we think hard about the future, that we think of it in the proper terms, and that we now and then temper our hope with caution.

As always, our ability to affect the future is far greater than our ability to know the future. But we do not need to know what is coming —or even to know what we want the future to bring—in order to know what we should hope to avoid. As Jonas put it, "what we must avoid at all costs is determined by what we must preserve at all costs."[17] Of course, it is also not always easy to know what we must preserve— what is crucially in need of defense and what, on the other hand, could be profitably traded for an improvement in our health, power, or wealth (or those of future generations). But one thing we surely must preserve, one thing we will certainly need regardless of what the future holds, is the capacity to rear and to educate future generations: the capacity for continuity. The quest for improvement and innovation is a force for great good, but it must not destroy the preconditions for its own efforts—the preconditions for the future. Chief among these preconditions is a continuity with the past, a link to roots. A society built on a profound cultural discontinuity—a severing of generations—would be like a cut flower, bright and beautiful for a time but no longer able to draw upon that which made it possible, and so unable to perpetuate itself.

To think of the future requires imagination; and to think of generations entering the world of the future requires a tremendous feat of imagination. In a strange way, it is precisely the most eager futurists in our contemporary politics who seem to lack the capacity for such feats of imagination, who see only themselves in the future, and fail to take account of the need to bring up those who will travel there, and those who will be born along the way. Responsible futurism requires that we imagine a world without us in it, and that we care about it. If the only way we can bring ourselves to care about the future is to make sure that we live forever, then we have little hope of doing the future much good.

The needs of future generations, just like those of past and present ones, extend beyond health and wealth and comfort. If they are to live well, and to raise those who follow them to live well, they must aspire to greater things. Life, liberty, and the pursuit of happiness are the minimal standards for decent living, not the highest ends of man. They are critical, but we cannot rest satisfied with them. We need larger aims, and the future will too.

Imagining the future through the lens of innovation leads us to believe that the most important challenge we will face in the future is steadily improving the material conditions of human life by steadily improving upon human understanding and power. Meeting that challenge requires individual freedom to innovate, and this must not be constrained for the sake of vague concerns about unpredictable consequences. But imagining the future through the lens of generations leads us to believe that the most important challenge we will face in the future is also the most important challenge we face today and have always faced in the past: the challenge of bringing up those who are new to the world. That challenge requires some basic prerequisites that must not be innovated out of existence. This, most fundamentally and clearly, is the difference of opinion at the heart of our science debates.

The difficulty is that both lenses show us something true about the future, and both also put us at risk of mistaking the present for the future—either by failing to imagine genuine progress, or by failing to imagine a world without ourselves in it. We are left to decide how to balance the lessons of these two competing anthropologies, for our sake and for the sake of the future. Our ongoing debates over questions of science and politics are an effort to seek just that balance, far more than they are really arguments about particular technologies.

Of the two competing visions, the anthropology of generations offers us a fuller and more recognizable account of the truth of the human condition. But it surely is not simply right, and if we are to secure the preconditions for progress, we must remember that we do this because progress is good for us and important, and not because we simply wish to preserve the world we have known. We must be careful, in tending our intuitions and hopes, to weed out simple reactionism, and to avoid the misguided desire for a wholesale recovery of the past. The past was not as good as we think we remember it was.

Instead, what we risk losing, and what we might want to recover, is something more like the past's way of thinking about the future. One of the most monumental innovations of the modern age—in science and society alike—has been a way of thinking about the future that ignores, and so leaves little room for, the future generations who will have to live with the consequences of our actions. This is one innovation we will have to resist in order to make truly responsible progress possible. The recovery of an older way of imagining the future need not be reactionary. It is not about pining for the past as much as it is about admiring the future made for us by those who came before, and seeking to build one no less admirable for those who will come after.

Keeping in mind the burdens and blessings of natality and the peculiar responsibility that the present always has for the future offers the only way for us to make moral sense of the new possibilities opened up by the age of biotechnology. If we can do it well, we will be better able not only to preserve and improve our moral traditions and to confer to our children an implicit sense of human dignity and human excellence, but also to preserve and to strengthen the preconditions for liberal and libertarian virtues and freedoms. The defenders of each must appreciate the importance of both.

Our science debates allow us to see more clearly than usual the disagreement between these two camps of our politics about how to envision the future. And having seen it, we can appreciate, too, that this dispute in fact underlies a great deal of our political life, and explains something crucial about the right-left divide. We can therefore now turn back to the science debates, with this knowledge in hand, and consider anew how the right and the left deal with science. We have seen what the left—the party of progress and innovation— seeks to achieve, but we should consider the dangers and the contradictions that quest presents to the left and to America generally. And

we have seen what the right—the party of continuity and of generations—seeks to defend. But that defense, too, carries grave risks and
complications. These twists and dangers in the struggle for the future
are where we now turn.

CHAPTER FIVE

Science and the Left

THE ANTHROPOLOGY OF innovation, which makes its home mostly (though not exclusively) in parts of the American left, seeks to use science to advance its cause. The right, meanwhile, with its eyes on continuity and generations, has at times a more ambiguous view of certain kinds of scientific advances. And yet in our science debates, for reasons both practical and philosophical, the left seems often to make its own case in a strikingly defensive tone. Indeed, a casual observer of American politics in recent years could be forgiven for imagining that the legitimacy of scientific inquiry and empirical knowledge are under assault by the right, and that the left has mounted a heroic defense. Science is constantly on the lips of Democratic politicians and liberal activists, and is generally treated by them as a vulnerable and precious inheritance being pillaged by Neanderthals.

"For six and a half years under President Bush," Senator Hillary Clinton told an audience in October 2007, "it has been open season on open inquiry."[1] Senator Edward Kennedy, in an April 2007 speech at the Massachusetts Institute of Technology, bemoaned the many ways in which "the truth is taking a beating" under conservative influence in Washington.[2] One popular recent book on the subject is entitled *The Republican War on Science*; another, by Nobel laureate and former vice president Al Gore, is called *The Assault on Reason.*

But beneath these grave accusations, it turns out, are some remarkably flimsy grievances, most of which seem to amount to political disputes about policy questions in which science plays a role. Ethical disagreements over the destruction of embryos for research are described instead as a conflict between science and ignorant theology. Differing judgments about the proper role of government in sex education in schools are painted as a quarrel between objective public health and medieval prudishness. A dispute about the prudential wisdom of a variety of energy policy alternatives is depicted as a clash of simple scientific facts against willful ignorance and greed. And the countless minor personnel and policy decisions that always shape the day-to-day operations of the federal executive branch are pored over in an effort to reveal a nefarious pattern of retrograde anti-rational obscurantism. George W. Bush's science advisor, it seems, was given an office a little further from the Oval Office than his predecessors had, and a member of an FDA advisory board once wrote a book about his religious conversion.

The American right has no desire to declare a war on science, and nothing it has done in recent years could reasonably suggest otherwise. The left's quixotic defensive campaign against an imaginary enemy therefore has little to tell us about American conservatives— who, of course, as we have begun to see, *do* have a complex relationship with science, though it is not the one the left seeks to describe.

But if this notion of a "war on science" tells us little about the right, it does tell us something important about the American left and its self-understanding. That liberals take attacks against their own political preferences to be attacks against science helps us see the degree to which they identify themselves—their ideals, their means, their ends, their cause, and their culture—with the modern scientific enterprise. New Mexico Governor Bill Richardson seemed to speak for many when, in a speech in the course of his ill-fated campaign for the 2008 Democratic presidential nomination, he called upon

Democrats to make theirs "the party of science and technology."[3] This is a more positive (not to say less paranoid) way of expressing the deep connection between science—understood both as a way of knowing and a means to doing—and the agenda of liberalism and progressivism.

Putting aside all the loose talk of a Republican assault on reason, this simpler point does ring true: There is indeed a deep and well-established kinship between science and the left, one that reaches to the earliest days of modern science and politics and has grown stronger with time. Even though they go astray in caricaturing conservatives as anti-science Luddites, American liberals and progressives are not mistaken to think of themselves as the party of science, for some of the reasons we have already seen. They do, however, tend to focus on only a few elements and consequences of that connection, and to look past some deep and complicated problems in their much-valued relationship with science. The profound ties that bind science and the left can teach us a great deal about both.

THE PARTY OF SCIENCE

Every democracy in the modern age has seen its politics divided into recognizable camps of progressives and conservatives—a party of radical reform and social revolution and a party of tradition and social stability—and in the last chapter we got a sense of why that is, and brought out part of what lies beneath the division. The split has never been a clean one, of course, and in America in particular it has been complicated by the nation's liberal tradition (so that the conservative party often defends classically liberal ideals while the progressive party seeks to push beyond them). But nuanced though it always is, this divide has been a defining feature of the political life of the West in the last two centuries. It has been with us roughly since the time of the French Revolution—which indeed gave us the terms "left" and "right" for the two great streams of political instincts and attitudes.

It is not unfair to suggest that the right emerged in response to the left, as the anti-traditional theory and practice of the French Revolution provoked a powerful reaction in defense of the continuity of a political order built to suit human nature and tested and tried through generations of practice and reform. The left, however, did not emerge in response to the right. It emerged in response to a new set of ideas and intellectual possibilities that burst onto the European scene in the seventeenth and eighteenth centuries—ideas and possibilities that we now think of as modern scientific thought.

It is difficult for us today to fathom the extent of the excitement, optimism, and enthusiasm set loose upon Europe by the emergence of modern science in the seventeenth century. As its teachings and its applications spread (and particularly in the wake of Isaac Newton's extraordinary reformulations of natural phenomena in mathematical terms), their potential seemed boundless.

To the lay enthusiasts of the new science (Voltaire, for instance, who published numerous popularizations of Newton's ideas and began a kind of "cult of Newton" in French intellectual circles), its great promise was not only in its power to explain nature, but in its capacity to offer an alternative to tradition, and especially religious tradition.

By offering both new answers and a powerful new way to seek more answers to the mysteries of nature, the methods of modern science seemed to promise a direct path to knowledge that did not depend upon archaic tradition and faith and did not answer to the institutions of the church or the state. Using empirical observation properly assessed by straightforward scientific methods based in reason, individuals could abandon all they were told by authority and seek direct unmediated knowledge on their own. Knowledge, René Descartes and his followers asserted, could begin from scratch and build itself on firm foundations. The allure of starting from scratch in political life as well soon became hard to resist.

Descartes' and Newton's powerful new approach also introduced several other appealing methodological innovations. It worked by breaking large phenomena down to their constituent parts and understanding the whole as a function of parts, not the other way around (a mechanical, rather than organic, conception of nature). And it rejected teleology as well, and so sought to understand the objects of nature not by the ends toward which they were said to be oriented, but by the beginnings from which they appeared to emerge.

The power of these new and effective methods of knowing cannot be overstated. They were bound to burst beyond merely scientific thought—and in the realm of political thought, they quickly gave shape to a fervent anti-traditionalism, and to a thoroughgoing mechanism and individualism, all of which would play a great part in the drama of the modern age.

On the one hand, and particularly in Britain, these ideas led to a rational new political philosophy based precisely on individualism, materialism, and a historical explanation of human affairs. In the beginning, Thomas Hobbes and John Locke argued, was the individual alone—a state of nature of equal unconnected particles in motion. These individuals, the atoms of society, eventually joined to address their material needs and their fear of death (which happen also to be the needs and fears addressed by the new science), and society is best understood as a function of those individuals, needs, and concerns, all accounted for by rational contract and individual will. This new politics, which we now think of as classical liberalism, made possible a thoroughgoing case for human equality and liberty, defensible by postulates grounded in individualist materialist premises and building up with geometric exactitude.

On the other hand, and particularly in France, these same ideas (including Locke's political translation of them) contributed to a powerful zeal to overthrow tradition and replace it with rational design. The *philosophes* who paved the way for the French Revolution were not only intensely anti-traditional and anti-religious, they were also in many cases able students (and in a few cases, like Jean le Rond D'Alembert, even great innovators) of the new science, and not by coincidence. Their political principles emerged from an effort to produce a fully rational model of political life from scratch, a model that began from individualism, and that understood liberty and equality in terms borrowed directly from the parlance of the scientists. Science, for them, was a profoundly progressive liberating force, a great sword with which to slay mighty kings and priests, and a new path to ever-increasing knowledge that could only lead to ever-increasing freedom. It all depended on beginning every political action from the most basic possible material assumptions and reasoning from established premises to a conclusion—which in practice had to mean rejecting all that existed simply by habit, and beginning anew. Traditional institutions that had long endured could only be defensible if they, too, met this test of reason from scratch. The more thoroughly grounded in rational scientific thinking a society could be, the more legitimate and free it would be—and with time, the direction of progress was clear: away from tradition and toward rational mastery.

This is, of course, a crude brief history of several centuries of new ideas. And in some respects it even minimizes the degree of cross-pollination between the new politics and the new science—as the ideas of Machiavelli fed those of Bacon and Descartes, and theirs in turn gave rise to a liberal and a revolutionary politics. But when the modern left was born, largely in the great crucible of the French Revolution, these ingredients had long been combining, and the new way of thinking—anti-traditional, egalitarian, liberationist, progressive, highly rationalized, and always forward-looking—was bound up at its birth with modern science. The material aims of science suited the aims of the left, and the intellectual means of science were the preferred means of the left. Above all, the progressive vision of ever-growing knowledge, and with it ever-growing power (and therefore ever-growing freedom), is the common legacy of both.

The immense enthusiasm unleashed upon the world of ideas by the scientific revolution was a wave of progressivism—not just in science but also in politics. Its central assertion was that the future would be very different from the past, because the past was rooted in error and prejudice while the future would have at its disposal a new oracle of genuine truth. It was a revolutionary ethic of discontinuity the likes of which had never before been seen in human history.

That vision, of course, has always had an element of naïve utopianism to it, especially in its overestimation of the power of knowledge to liberate and an underestimation of the need for traditional restraints on human willfulness. This, too, has been a common legacy of science and the left. "The condition of the world being materially changed by the influence of science and commerce," Thomas Paine wrote in 1782, "it is put into a fitness not only to admit of, but to desire, an extension of civilization. The principal and almost only remaining enemy it now has to encounter is prejudice; for it is evidently the interest of mankind to agree and make the best of life."[4] A lovely foolishness, of course, and still very much a strand of left-wing thinking, in both its loveliness and its foolishness.

As the left has advanced from its birth in the French Revolution, it has kept its eye on the value of science as a way of thinking and a way of doing. Its excesses—like the gruesome experiment in applied social science called communism—have drawn on science and pseudoscience. And its successes—like progressive social and public health reforms—have often done the same.

For modern progressives, science as a method of knowing has been especially important. John Dewey, perhaps the greatest philosopher of American progressivism, argued that the scientific method, "a shorthand designation for the great and ever-growing methods of observation, experiment, and reflective reasoning," held the key to the solutions to social problems—and that these problems persist only because scientific methods of knowledge and reasoning have yet to be applied to them. "Science," Dewey wrote in 1946, "bears exactly the same relation to the progress of culture as to the affairs acknowledged to be technological (like the state of invention in the case, say, of tools and machinery, or the progress reached in the arts, say, the medical)." He continued:

A considerable part of the remediable evils of present life are due to the state of imbalance of scientific method with respect to its application to physical facts on one side and to specifically human facts on the other side ... the most direct and effective way out of these evils is steady and systematic effort to develop that effective intelligence named scientific method in the case of human transactions.[5]

The great contribution of science, in this sense, is not so much in the technologies it makes possible, as in the modes and methods it has developed for understanding problems and conceiving solutions. Science is a path to knowledge, and for American progressives, scientific knowledge is the path to social progress.

Both as action and as knowledge, then, science has been a source of inspiration for progressives and for liberals, and its advancement has been one of their great causes. That does not mean that science captures all there is to know about the left. Far from it. The left has always had a deeply romantic and even anti-rationalist side too, reaching back almost as far as its scientism. But in its basic view of knowledge, power, nature, and man, the left owes much to science. And in the causes it chooses to advance in our time, it often looks to scientific thought and practice for guidance. In its most essential disagreements with the right—in particular, about tradition—the vision defended by the left is also a vision of scientific progress.

The left is therefore generally justified in thinking of itself as the party of science. But for all its advantages, this relationship is not a simple matter. It is subject to some important and complicated tensions, which are emerging with special force in our own day.

POWER AND PROGRESS

The great original appeal of the scientific enterprise was its potential to empower man over nature. Francis Bacon set out the conquest of nature as his aim. René Descartes sought to make human beings "masters and possessors of nature."[6] And the scientific community they helped to found has since continued to pursue these twin objectives: expanding human power and conquering nature.

But for the modern left, each of these key aims of modern science has grown deeply problematic. To begin with, over the past century the left has come to take a rather complicated view of power. It has become highly suspicious of certain kinds of power: the power of nations, of corporations, of the rich over the poor, of man over nature (or as it has been renamed, to make it passive, "the environment").

Much of this change took place in the course of the twentieth century—a time of previously unimaginable inhumanity and villainy. Shaken by examples of power run amok, and by exposure to and interaction with postmodernism (with its excessive and blinding obsession with power), many on the left became opponents of power as such, in ways that earlier progressives had decidedly not been. This is evident in the ethic of the environmental movement, in progressive views of foreign policy and economics, and in the general tenor of the left.

But this suspicion of power seems not to have made much headway in the left's views about the two most powerful institutions of the age: the state and science. This is easier to explain when it comes to the state, which American liberals and progressives have taken to be the essential institution of social solidarity, political expression, material improvement, and justice. The ideology of the left is centered upon a proper employment of the power of the state, and so the left is naturally disinclined to turn against the use of such power.

But blindness to the power of science is a more perplexing quandary, and one not yet seriously faced by the left. To the extent they have sought to confront it (as in the mainstream of bioethics), it has been to rationalize scientific power, and enable its application within the liberal boundaries of consent and safety. But there is much more than that for the left to think about. Science (as the true postmodernists know) is the foremost font of modern power, and the underlying source of almost all the expressions and incarnations of power the left does find troubling: industrial power, corporate power, military power, imperial power, and especially human power over the natural world.

Indeed, it is in the arena of environmentalism, more than anywhere else, that this blind spot of the party of science is most pronounced. There, the left's problem with power and the left's problem with conquering nature become one—yet the role science plays in making both possible has never come front and center.

THE CONQUEST OF NATURE

In the past three decades, environmentalism has become a fully integrated component of the worldview of the American left, the party of science. But the perspective of environmentalism could hardly be more different than that of modern science on the questions of nature, power, progress, and man.

Modern science is grounded in a particular view of nature, both material and moral. The natural world, thought the fathers of science, is matter in motion; it is best understood by being pulled apart into its constituent forces and pieces and experimented upon under duress. "The nature of things betrays itself more readily under the vexations of art than in its natural freedom," Bacon argued, because nature is not a whole but a sum of parts, and is not moved by a purpose, but driven by discrete causes alone.[7] Nature, moreover, is the chief constraint on human power and human comfort, and the extension of the empire of man over nature is a noble and necessary goal. For too long, human beings had been subject to the whims of nature and chance, but by coming to know the workings of nature, we could master it, both removing natural obstacles and constructing artificial advantages for ourselves. "Nature, to be commanded," Bacon wrote, "must be obeyed," so the purpose of the new natural science was to learn nature's ways so as to overcome them.[8] This desire for knowledge of and power over nature was not power-hunger, it was humanitarianism. Nature, cold and cruel, oppresses man at every turn, and bold human action is needed in response. Science arose to meet that need.

If you had to devise a complete opposite to this scientific view of nature, a mirror image in essentially every respect, you would probably end up with roughly the notion of nature that gives shape to the modern environmentalist ethic. Nature in this view is, to begin with, a complete and ordered system, to be understood in whole and not in part. "When we try to pick out anything by itself, we find it hitched to everything else in the universe," wrote John Muir, a founder of modern environmentalism.[9] Far from conquering and manipulating nature for his benefit, moreover, man must be careful and humble enough to tread gently upon it, and respect the integrity (and even the beauty) of its wholeness. We are to stand in awe before nature, and never to overestimate our ability to overcome it or underestimate our ability to harm it (and with it ourselves). "We have forgotten how to be good guests, how to walk lightly on the earth as its other creatures do," wrote the great British environmentalist Barbara Ward in her 1972 book *Only One Earth*.[10]

Taken to the extreme, this approach turns the scientific view of nature on its head, and looks at man as an oppressor of the natural world instead of the other way around. The title of one popular recent book, for instance, imagines the peace and beauty of *The World Without Us*. "How would the rest of nature respond if it were suddenly relieved of the relentless pressures we heap on it and our fellow organisms?" the author asks.

> How soon would, or could, the climate return to where it was before we fired up all our engines? How long would it take to recover lost ground and restore Eden to the way it must have gleamed and smelled the day before Adam, or *Homo habilis*, appeared? Could nature ever obliterate all our traces?[11]

Not all environmentalism indulges in such anti-humanism, to be sure. But in all of its forms, the environmentalist ethic calls for a science of beholding nature, not of mastering it. Far from nature the oppressor, this new vision sees nature as a precious, vulnerable, and almost benevolent passive environment, held in careful balance, and under siege by human action and human power. This view of nature calls for human restraint and humility—and for diminished expectations of human power and potential.

The environmental movement is, in this sense, not a natural fit for the progressive and forward-looking mentality of the left. Indeed, in many important respects environmentalism is deeply conservative. It takes no great feat of logic to show that conservation is conservative, of course, but the conservatism of the environmental movement runs far deeper than that. The movement seeks to preserve a given balance which we did not create, are not capable of fully understanding, and should not delude ourselves into imagining we can much improve— in other words, its attitude toward nature is much like the attitude of conservatism toward society.

Moreover, contemporary environmentalism is deeply moralistic. It speaks of duties and responsibilities, of curbing arrogance and vice. As Charles T. Rubin puts it in his insightful 1994 book *The Green Crusade*, "environmentalism is the temperance movement of our time," albeit largely devoid of the religious convictions that moved those prior progressives.[12] Think "addicted to oil." It is a movement stirred by moralism to reform a prominent human excess, and driven by the hope that this reform will improve almost everything about life. As Al Gore put it when he was given a Nobel Peace Prize for his environmentalist efforts, "the climate crisis is not a political issue; it is a moral and spiritual challenge to all of humanity."[13]

Indeed, writ large, the environmental movement aims to repeal the modern way of life. At its most ambitious, it seeks to curb industrialism and consumerism, to make the human experience less artificial and more "authentic" (or, to employ the favored buzzword of the day, "organic"), to emphasize the simple and the local, to reduce the scale of human ambition. This describes a brand of conservatism too conservative even for the American right, and one that is deeply at odds with the ethic of rationalization and scientific improvement and progress.

Some elements of this approach are not entirely new to the left, at home or abroad. The yearning for authenticity and simplicity, the revulsion at power, and the skepticism of technology and systematic knowledge have been elements of what came to be known as "the new left" in the late 1960s, and to some extent had characterized progressive politics for far longer too. They have had a lot to do with shaping the ideology of left-wing parties throughout the West. But the manifestation of this approach in the modern environmentalist movement is far more prominent, more powerful, and, for the left, more complicated than any other.

It is prominent and powerful because environmentalism, and particularly concern with global climate change over the past decade or so, has come to play an astonishingly central role in the politics of the West. In a time when Iran is reportedly pursuing nuclear weapons, North Korea is violating international agreements, the future of Iraq remains uncertain, genocide persists in Sudan, and countless other crises threaten the peace of the world, Ban Ki-moon, upon taking his post as Secretary General of the United Nations in 2007, listed climate change as his top priority. "The danger posed by war to all of humanity and to our planet," he said, "is at least matched by the climate crisis and global warming."[14] European Commission President José Barroso has argued that climate change must be the European Union's top priority as well.[15] German Chancellor Angela Merkel has called it "humanity's greatest challenge."[16] Even stipulating the basic facts regarding global climate change, this kind of attitude is surely absurd.

There is no question that for some, especially in Europe, the obsession with climate change is a way to avoid thinking about serious geopolitical problems, particularly the threat of radical Islam. Rather than marshalling modernity to defend itself, this obsession allows Western elites to persist in a silly and feckless pseudo-moralism. Instead of looking to America for leadership and protection, it allows them to blame America for its strength and its confidence. And for some on the left, too, the obsession is a way to stir up the kind of crisis atmosphere necessary for some pet causes and ideas to become politically plausible. But whatever the reason, environmentalism, and with it a worldview deeply at odds with that behind the scientific enterprise, has come to play a pivotal role in the thinking of the left.

So far, the American left has managed mostly to ignore this difficulty, and to treat environmentalism as a cause of the party of science. An ongoing dispute about the basic facts and figures of global warming has made this easier by putting science and environmentalism on the same side for a time. But as that argument subsides, and attention turns to the causes of environmental degradation and to possible solutions, the fissure between science and environmentalism will be harder to ignore. An American environmentalism newly empowered by a decades-long debate that put it front and center on the agenda of the cultural and political left may come to resemble the European Green movement, which shares many of the attitudes of American progressives, but which does not view itself by any means as a party of science. Indeed, the Greens in Europe have been at the leading edge of nearly every contemporary effort to curb the power and the reach of science, most notably biotechnology—from bans on human cloning to prohibitions against genetically modified foods. But in America, the left has yet to confront this glaring complication in its claim to the mantle of the party of science. Science, it turns out, is behind much of what troubles and worries the left.

CREATED EQUAL

To the extent that Americans have pursued any serious limits on science in recent years, they have done so not as Greens concerned for the integrity of nature, but as conservatives concerned for the equality and dignity of man. And the most politically potent of these efforts have been grounded not in human dignity—a crucial concept, though one still sorely in need of intellectual refinement—but in human equality, or equal humanity, largely understood.

The defense of dignity is a defense of the stature of man, and a reaction against efforts to demean or lower him. Some concerns about science take this form, as for instance when critics worry about enhancement technologies that could undermine the meaning of human performance, or about the potential of human cloning to distort family relations and confound human identity.

But as these tend to be futuristic worries, and as dignity so understood is something of an aristocratic notion, these concerns have not been key to the bioethics of the right. Most conservative critics of science (and particularly of biotechnology) are worried for human equality—and indeed very often when they speak of human dignity, they actually have equality in mind.

Human equality is, of course, the great American ideal, inscribed upon the nation's birth certificate as a self-evident truth. Equality has also been a defining cause of the left, from the French revolutionaries marching under the banner of "*liberté, égalité, fraternité,*" to the movements for civil rights and equal treatment in America, to the commitment to democracy and the economic theories that continue to shape the politics of liberals and progressives. The party of science has, from the beginning, also been a party of equality. And at the dawn of the modern left, the advocates of science and of equality believed the two great projects would advance together. "The general spread of the light of science," Thomas Jefferson wrote in his last days, "has already laid open to every view the palpable truth that the mass of mankind has not been born with saddles on their backs, nor a favored few booted and spurred, ready to ride them legitimately, by the grace of God." [17] Science, he believed, would simply demonstrate human equality. But it has not worked out that way, and today modern science poses greater and deeper challenges to our belief in human equality than any other force in modern American life. Science exacerbates key difficulties with equality, and equality points to the limits of the scientific worldview—although the left, which seeks to advocate both, has not yet fully come to see this.

When the left is critical of science and technology on egalitarian grounds, its concerns tend to focus on unequal access to benefits. The emergence of new technologies, it is argued, contributes to inequality by creating new haves and have-nots. This has been a common concern from the beginning of the industrial age, but the evidence of history suggests it is not well founded. New technologies can surely exacerbate some existing inequalities, but they can also help ameliorate them, and in general they are not the cause of novel social inequalities. The fruits of technology, especially in democratic societies, have made their way to all levels of society fairly quickly. We no longer hear much about "the digital divide" that was much on the lips of social critics of computer technology just a decade ago, for instance, and it is unlikely that new biotechnological advances will create lasting distributive inequalities either.

The trouble science poses for egalitarianism runs much deeper than that: It involves on the one hand a weakening of the case for human equality, and on the other a positing of ends and purposes taken to be higher and more important than equality.

Science, simply put, cannot account for human equality, and does not offer reasons to believe we are all equal. Science measures our material and animal qualities, and it finds them to be patently unequal. We are, after all, obviously not all equally large or small, tall or short, strong or weak, healthy or ill. We are born physically and mentally unequal, and always remain so. To examine only our animal qualities is surely to conclude that we are far from one another's equals. And so to assume that there is nothing more to us than our animal qualities (as the modern scientific outlook does) is to assume inequality is the human condition.

Yet it is in precisely the ways in which we are more than animals that any serious case for equality is grounded; to imagine that no such ways exist is to assert that no such case could be valid. The closer the left aligns itself with the ideology of modern science (taking, for instance, all human actions and beliefs to be mere functions of neural biochemistry) the further it seems to distance itself from any sensible case for egalitarianism.

The case for human equality does not require a rejection of empirical science, but it does appear to depend on some sense of the limits of science's reach, and on the view that some elements of the human experience are best understood in something other than scientific terms. Equality, of course, does not mean sameness or identity. We are not all the same even if we are all each other's equals. As understood in modern times, rather, it describes a minimal but essential common humanity, important especially in establishing restraints on what those who bear it in common may do to one another. It is possible, in fact, to simply replace the comparative word "equality" with the non-comparative word "humanity" in most instances, and not lose much in the trade. Our humanity—rather than any particular ability or capacity—is that in which we are equal. It is a very modest claim, but it goes a long way, because in the modern age we have built a great edifice on the grounds of this equality.

There are, in fact, several fairly distinct types of modern arguments for equality, each best understood on its own terms. Some are far more vulnerable than others to the scientific case against equality.

One common and powerful case for human equality begins with the Judeo-Christian notion that human beings are made in the image of God. We are equal to every human being and unlike every other creature, this transcendent view asserts, as bearers of that image and that relation to the divine. This is also, for some, an argument for human dignity: equal dignity as creatures marked by divinity.

A related case for equality relies on the common created origins of human beings to assert a foundation for equal treatment in society. As the Declaration of Independence puts it, "all men are created equal" and "are endowed by their Creator with certain unalienable Rights." This approach makes equality itself the premise upon which all other political institutions and arrangements are constructed, and therefore does not so much make a case for equality as build all other cases upon the assumption (or self-evident truth) that all are equal. This is a less expressly religious egalitarian case: it posits God's creation of man as the source of human equality, but does not further rely on the divinity of that beginning. It relies upon the modern (and in some respects scientific) notion that origins explain everything. Since human beings are equal in their origin, they must also be equal in their social standing.

This view builds on the classical liberal case for equality, which is grounded in a theory of the origins of society. It posits a time before all society—a state of nature—and takes man's bearings by what might have been true of him in such a state. The state of nature, John Locke writes, is

> a state also of equality, wherein all the power and jurisdiction is reciprocal, no one having more than another; there being nothing more evident, than that creatures of the same species and rank, promiscuously born to all the same advantages of nature, and the use of the same faculties, should also be equal one amongst another without subordination or subjection.[18]

This equality is so evident because there are no pre-existing social connections to define different ranks. Human beings in this state are unconnected individuals, and so are all on the same level—whether equally covetous and miserable (as Thomas Hobbes would have it) or equally free and desirous of peace (as Locke would). Liberal egalitarianism, in this sense, is actually a function of liberal individualism. In fact, in his original draft of the Declaration of Independence, Thomas Jefferson asserted "that all Men are created equal and independent," not just equal. The origins of society are then in the rational agreement of unconnected equal individuals: equally men; equally endowed with rights to life, liberty, and property; and equally vulnerable to nature and their fellow men. All association follows from that premise.

The premise, however, is highly dubious. Liberalism's creation myth—that society's natural history begins with unconnected individuals—allows for an internally consistent case for equality, but it surely bears no connection to the actual history of humanity. As the philosopher David Hume put it, "men are necessarily born in a family-society at least."[19] Conservatives, beginning especially with Edmund

Burke, have been harsh critics of this terribly implausible liberal tale of beginnings. And in our time most on the left don't take it very seriously either.

Instead, some, following the lead of Immanuel Kant, have made a case for equality based on the rational capacity of the human person —a case that celebrates reason as worthy of respect, and shows regard for the rational animal; and therefore a case that values man for possessing a particular ability, rather than merely for being human. This is more of a case for rationalism than equality, though in practice it can build a foundation for significant protections of human rights, even if it cannot explain the source or nature of those rights.

But most liberals and progressives, to the extent they have thought through an argument for equality, have adopted egalitarianism as a means to justice, or more precisely to fairness, clinging to the ideal of equality because it is useful, or because it works, rather than because it is self-evidently true. This is not a bad reason to insist on equality, but it does make for a weak and very vulnerable egalitarianism. The exemplar of this approach is the political theorist John Rawls. Rawls takes for granted that the state of nature and the social contract are imaginary, and proceeds to ground his own theory even further from actual human experience. He asks readers to imagine themselves designing a society from scratch with knowledge of how politics works, but no knowledge of what position they themselves would occupy in that society. In that situation, he reasons, all of us would choose an egalitarian society with equal opportunity for all—just in case we found ourselves on the lower rungs of the social ladder. Such a society is therefore the most fair and just (indeed, Rawls argues, fairness is the very definition of justice), and offers a model of what real societies should strive for. In other words, we would do best to *pretend* that all men are created equal.

This tentative and purely functional commitment to equality— an egalitarianism of convenience—is exceedingly vulnerable to being undercut by science. Science not only provides empirical evidence against material human equality (which in itself of course does not undercut the case for equal humanity), but it also sometimes proposes means to material ends—to comfort, to wealth, to power, to health— that rely upon unequal treatment of human beings at the margins. The left's thin egalitarianism is ill-equipped to resist such an offer.

LIBERAL EUGENICS

American progressives have stumbled on this path before. In the late nineteenth and early twentieth centuries, the cause of material progress and scientific control, together with some crude misapplications of Darwinism, combined to form an energetic and progressive program of eugenics, beginning with public education toward selective breeding based on valued family traits, and culminating in a massive project of sterilization—including coercive sterilization laws in more than twenty states—of those found mentally or physically wanting. Nearly every prominent American progressive championed eugenics as an appropriate application of scientific knowledge to the nation's social ills. Herbert Croly, founder of *The New Republic*, argued in 1909 that to "improve human nature by the most effectual of all means—that is, by improving the methods whereby men and women are bred" would be crucial to social reform.[20] Margaret Sanger, the progressive activist and founder of Planned Parenthood, wrote in 1922 that "drastic and Spartan methods may be forced upon society if it continues complacently to encourage the chance and chaotic breeding that has resulted from our stupidly cruel sentimentalism."[21]

That "stupidly cruel sentimentalism" was, of course, American egalitarianism. Eugenics was most fundamentally a denial of human equality. By holding the quality of the gene pool above the equality of mankind, it forced a choice between science and equality, and most American progressives made the wrong choice. Eugenics was in time fatally tainted by its association with Nazi practices in the Second World War, though in some states eugenic laws and practices remained in effect well into the 1950s and even the early '60s. Most American progressives were not Nazi sympathizers by the time the war came, though, and so could legitimately distance themselves from the practices Americans found most abhorrent. This meant that the way eugenics went out of style actually protected progressives from fully coming to terms with their earlier commitment to scientific selection, and therefore with the tension between science and equality.

But such a reckoning, long put off, appears now to be nearing again. Today, a rather different sort of effort to apply control and selection over the next generation is emerging, in the form of a growing movement to test developing human embryos and fetuses for ailments and weaknesses (or even just the wrong sex), and to eliminate those found to bear them. The trend itself is undeniable. Ninety percent of Down syndrome pregnancies in America are aborted, for instance.[22] In Europe, according to one recent study, "40 percent of infants with any one of eleven main congenital disorders were aborted" between 1995 and 1999.[23]

Selection of embryos based on genetic traits (through what is called preimplantation genetic diagnosis) is becoming an increasingly routine element of in vitro fertilization treatment. And as the British IVF pioneer Robert Edwards put it in 1999: "Soon it will be a sin of parents to have a child that carries the heavy burden of genetic disease. We are entering a world where we have to consider the quality of our children."[24] This is the language of the new eugenics.

Its defenders argue that this "liberal eugenics" (as the British writer and advocate of such practices Johann Hari has dubbed it)[25] is fundamentally different from that of the early twentieth century because it is not coercive or state-mandated but instead is a matter of individual or family choice. It is also not based on race distinctions or assessments of intelligence or social class. It is often (though not always) carried out by parents, when they discover their child has a condition they believe would be a grave detriment to his or her welfare or happiness, or (less often) when they find out that their child is simply not the kind of child they want. "Much of the bad reputation of eugenics," write the liberal bioethicists Allen Buchanan, Dan W. Brock, Norman Daniels, and Daniel Wikler, "might be avoidable in a future eugenic program."[26]

These differences are certainly significant. But surely the most essential problem with the eugenics movement was not coercion or collectivism. It wasn't even the revolting notion of a duty to improve the race. The deepest and most significant contention of the progressive eugenicists—the one that made all the others possible—was that science had shown the principle of human equality to be unfounded, a view that then allowed them to use the authority of science to undermine our egalitarianism and our regard for the weakest members of our society.

Today, as then, belief in equality remains essential for much of the worldview and agenda of the left. But today, as then, the left finds itself ill equipped to defend that belief against this kind of assault. The egalitarian justice of John Rawls offers little help, and indeed Rawls himself made plain that his theory was compatible with eugenics. In his magnum opus, *A Theory of Justice*, Rawls argues that given his principles, each generation should be seen to owe the future a society with "the best genetic endowment," and that "thus, over time a society is to take steps to preserve the general level of natural abilities and to prevent diffusion of serious difficulties."[27] He is not specific as to just how this might be accomplished, but it is hard to conceive of an egalitarian means for achieving it. If equality is purely a means to an end, then a more effective means to that end will be hard to resist, and equality will be easily swept away as a needless obstruction.

SCIENCE VERSUS EQUALITY

This twofold challenge science poses to equality—dismissing it as unfounded on the one hand and condemning it as an obstacle to material improvement on the other—has crystallized in recent years in the heated public dispute over embryo research. The capacity of the left to stand firm on the ground of egalitarianism has been tested in that argument, and found badly wanting.

The stem cell debate, which began in the 1990s and which may soon begin to subside thanks to scientific developments that provide alternative sources of cells that do not require the destruction of embryos, has been mired in confusion from the start. In large part because of its connection to the (related but distinct) abortion debate, the simplest terms of the argument have not been well understood.

The debate is, to begin with, not about stem cell research, any more than an argument about the lethal extraction of livers from Chinese political prisoners would be a debate about organ transplantation. There are ethical and unethical ways to transplant organs, and the same is true for stem cell research. The question is to which category a particular technique—the destruction of living embryos for their cells—belongs.

The debate is also not about whether there ought to be ethical limits on science. Everyone agrees there should be strict limits when research involves human subjects. The question is whether those limits should apply to the case of human embryos. But that does not mean the stem cell debate is about "when human life begins." It is a simple and uncontroversial biological fact that a human life begins when an embryo is created. That embryo is human and it is alive; its human life will last until its death, whether that comes days after conception or many decades later in a nursing home, surrounded by children and grandchildren. All of us were once embryos, and none of us were sperm or egg. Our lives began when the gametes combined to form a new human being.

But the biological fact that a human life begins at conception does not by itself settle the ethical debate. The human embryo is a human organism, but is this being—microscopically small, with no self-awareness and little resemblance to us—a person, with a right to life?[28]

At its heart, then, when the biology and politics have been stipulated away, the stem cell debate is not about when human life begins but about whether every human life is equal. Is our equal humanity enough to merit even the most minimal protection and regard—the protection from being killed on purpose—or does our equal humanity in the end mean essentially nothing, while only some crucial capacities and abilities, which we possess to varying degrees in the course of our lives, can qualify us for protection from harm? Do we refrain from mistreating fellow human beings because of *what they can do,* or because of *what they are?* The circumstances that have forced this question—the ability to create a human embryo outside the body of a mother—have also put it in the most exaggerated form imaginable, but they do not change the question. It is true that the destruction of embryos for research might yield great medical benefits, and so could help us or our loved ones in a time of grave need. And yet it is also true that human embryos are human beings in the earliest stages of development. Which truth is more important?

The answer depends on one's view of the truth asserted as self-evident by the Declaration of Independence, and the science of embryo research has forced us to confront that question in the starkest possible way. Shall we treat the human embryo as less than human because it would be more useful to us dead than alive? Or shall we treat our equal humanity as a vital break on our ambitions, even when it comes at a price? The Kantian case for equality—which respects the bearer of reason—shows little regard for this human being at the fragile and mindless beginnings of life (or indeed for those at its frail and fallen end). Meanwhile, the thin functional case for equality finds it hard to ignore the greater use to which embryos might be put if only we disregard their humanity. The prospects of embryo research have caused liberal egalitarianism to come under attack by liberal humanitarianism. And the left has chosen to side with the latter, forgetting how profoundly it depends upon the former.

The American left seeks to be both the party of science and the party of equality. But in the coming years, as the biotechnology revolution progresses, it will increasingly be forced to confront the powerful tension between these two aspirations. In some instances, as apparently in the stem cell debate, it will be possible to avert the difficult choice (though even doing this will require a commitment to equality sufficient to elicit the necessary scientific creativity for a solution). In other instances, a choice will be called for, and the character of the left in the years to come will heavily depend upon the choice it makes.

To choose well, the American left will need first to understand that a choice is even needed at all—that this tension exists between the ideals of progressives and the ideology of science.

THE UNEASY ALLIANCE

Mastery of chance and of the given world is the deepest progressive longing, and so it is not surprising to find progressives on the side of science. But that same desire for mastery, and especially the rejection of the given, is also a denial of respect for equality and ecology, which progressives continue to claim among their highest ideals. Both ideals rely upon the presence of some unmastered mystery—some order beyond our grasping reach. A turning away from that humbling mystery, and toward unbounded will, is the inevitable (and indeed intentional) consequence of the progress of the modern scientific enterprise. That progress brings with it immense benefits, but if left to itself it threatens a great deal as well, including much that is of importance to the left.

Meanwhile, the left has also adopted an easygoing relativism about moral and cultural questions, so that science has come to be seen as the only source of objective knowledge—of knowledge equally true everywhere and all the time. Science thus cannot help but be elevated to an almost spiritual level, and to exercise an even more powerful pull on the thought and the politics and the imagination of the left, exacerbating the tensions inherent in the worldview of the party of science.

Recent political enthusiasms have aggravated these tensions all the more. The desire to win the stem cell debate (which proceeded under the shadow of the even more heated abortion debate) has driven the left closer to a rejection of equality than it might otherwise have been inclined to move. And the dispute regarding global warming has tied the left to an environmentalism that is in many respects a very strange bedfellow for liberalism. In the throes of political combat, however, these tensions have been obscured, and an imaginary larger fight for science—the enthusiastic counter-attack against a nonexistent "assault on reason"—has further helped to keep them hidden. But they will not remain hidden for long. In defense of science, the left has turned on itself, and forced to the surface some serious questions about its principles and priorities.

The answer, as ever, is moderation. The American left, like the American right, must understand science as a human endeavor with ethical purposes and practical limits, one which must be kept within certain boundaries by a self-governing people. In failing to observe and to enforce those boundaries, the left threatens its own greatest assets, and exacerbates tensions at the foundations of American political life. To make the most of the benefits scientific advancement can bring us, we must be alert to the risks it may pose. That awareness is endangered by the closing of the gap between science and the left—and the danger is greatest for the left itself.

It can be very difficult for today's liberals to see these dangers, because they touch such sensitive elements of the left's own view of the world. Some American conservatives are therefore far more keenly alert to these kinds of challenges and, in an effort to protect the prerequisites for cultural continuity as they understand them, are eager to address the excesses of science in the public square. But just as the left's attachment to science in the effort to secure the means of progress puts at risk some of the very ends the left considers most crucial, so the right's defensive efforts carry profound and paradoxical risks. In its attitude toward science, the right, just like the left, also might endanger precisely what it seeks most to protect. That is where we turn next.

CHAPTER SIX

Science and the Right

THE RIGHT SEEKS to protect treasured traditions as prerequisites for progress, while the left seeks to overcome them in the cause of progress. Science, as we have seen, far more readily defeats than defends tradition. It therefore sometimes poses uniquely serious dangers to the causes of the right, and demands a uniquely vigorous response from conservatives. That very vigor, however, carries grave risks: It can lead to overreaction, and it can even threaten to eviscerate the very institutions and traditions conservatives seek to defend.

The first of these dangers is the more prosaic, and also the more manageable. But it is a serious concern. The right can easily overreact to the dangers science poses to the means of cultural continuity. Conservatives risk equating science and the left—just as the left does—and so establishing themselves as enemies of empirical knowledge or of material progress, rather than defenders of the means of moral progress. These roles are quite distinct, and in confusing them the right threatens to fall into simple reactionism.

There are hints of this attitude, for instance, in the excesses that sometimes present themselves in the debate over the teaching of evolution in public schools. That argument, which is perhaps the oldest of the ongoing science debates, is often miscast as a dispute about scientific freedom. But it is really a debate about the authority of science as a source of knowledge of nature on the one hand, and about the significance of our knowledge of nature to our understanding of mankind (and so about the range of questions and answers that ought to define that understanding) on the other. To the extent that conservatives in this debate argue that local communities should have the right to teach their children what they wish, and that there is more to know about human origins than simply the scientific story, they are on reasonably solid ground. Our understanding of our common origins has a great deal to do with our sense of who and what we are as human beings, and to argue that only a purely material description of only our biological development has a place in the education of our children is to claim rather a lot. I would want my children to be taught the facts of evolution, but I am not troubled if my neighbor would rather that his children were taught a different kind of story of origins addressed to a different realm of human knowledge, with a different point to make. Nothing crucially common to us is lost or damaged when we make these choices; and nothing about being an American requires one to know a particular set of facts about natural history, let alone to accept the dubious proposition that scientific knowledge is the only knowledge relevant to the question of human beginnings, and so of human nature.

But when opponents of teaching evolution try to contend that evolution is simply false even as a merely scientific description of the natural history of life, they are in a very weak position, and arguing on turf they cannot seriously claim or defend. In those instances, they push too far in asserting limits to the reach and relevance of science, and seek to deny material facts because they take them to entail certain moral conclusions. They in effect adopt the very scientific determinism they are trying to combat; and they accept the proposition that the claims of evolution are in direct competition with the claims of Biblical religion or traditional morality, when in fact each offers answers to a different set of questions altogether. Adopting an exaggerated notion of the significance of the material data uncovered by science, they pursue excessive confines on science's reach—seeking not only to prevent it from overrunning its bounds, but also to deny its validity within those bounds. Evolution, properly understood, neither confirms nor denies human dignity. Its defenders surely sometimes go too far in asserting its consequences for our understanding of mankind, and their excesses should be answered; but to answer them with an equal and opposite excess, which denies the material facts and not only their supposed moral implications, is to replace one error with another. We should seek to understand evolution properly and avoid exaggerating its repercussions, not to deny it as far as it goes.

There are certainly other instances of conservative overreaction to science, and most involve similar examples of a misplaced determinism. A few conservatives, in fact, consider scientific method and technology inherently tainted and tainting, and worry that all of modern life is hopelessly corrupted by its contact with them. Seeking authenticity and simplicity, these conservatives—an assortment of agrarian traditionalists and crunchy urbanites—echo some of the romanticism of the green movement, and even of the postmodern technophobes. They seek to protect the roots of our way of life, but sometimes go too far in judging their vulnerability to science and not far enough in judging the value of science to that very way of life.

The risk of such overreaction is real, but with the exception of the evolution debate it is also fairly well contained. That debate takes up the question of the ability of science to spread its teachings without constraint, rather than to study and to act upon the world, and so treats science as a vehicle of culture above all. This puts science in a very unusual position, and (because culture is so crucial for conservatives) also poses a unique enticement to overreaction for the right. It requires special care from both sides, but does not often receive it from either. Beyond that quite unique debate, however, the right is fairly well alert to the perils of overreacting to science, and generally avoids them. For the most part, the American right has not been hostile to science. Indeed, on the whole the right has been quite friendly

www.encounterbooks.com
Please add me to your mailing list.

Name

Company

Address

City, State, Zip

E-mail

Book Title

ENCOUNTER BOOKS

900 Broadway

New York, New York 10003-1239

to the cause of science and technology, except in quite rare instances when, as we have seen, science seems genuinely to pose a threat to essential practices and institutions of cultural transmission.

It is the second risk inherent in the vigorous conservative response to science that raises truly serious difficulties. This more obscure but more grave peril emerges precisely when conservatives seek to define the difference between dangerous and beneficial scientific advances— or in other words, to define clearly what it is they seek to defend. The very process of definition, it turns out, can be destructive of the ideals and institutions being defended. This is a strange conundrum, which modern conservatism has wrestled with since its birth, but which emerges with particular force in the science (and again especially the biotechnology) debates. It is the special quandary of a conservative bioethics.

CONSERVATIVE BIOETHICS

The bioethics movement in America may be said to have begun in earnest in the late 1960s, when the Hastings Center was created as the first bioethics think tank. Its task was to advance the study of the ethics of biology and medicine, and to examine the moral and social significance of new developments in genetics, psychopharmacology, reproductive medicine, and other new frontiers of biological science. The movement has since grown by leaps and bounds, and bioethics has developed into a profession, if not an industry.

Some American conservatives have long shared the concerns that animate bioethics. The pro-life movement has always worried deeply about the treatment of the unborn by scientists and doctors, and many conservatives have through the years been interested in various issues surrounding medical ethics, illicit drug use, assisted suicide, and other social and cultural matters that have much to do with modern science. But it was not until fairly recently that bioethics emerged as a general and prominent category of concern for the American right.

That concern has been particularly influenced by worries about what has been dubbed the "Brave New World." This allusion to Aldous Huxley's famous book hints at a vision of a world reshaped by biotechnology: procreation replaced by manufacture, the pursuit of happiness replaced by drugs, and human nature remade into something lower and shallower, more easily satisfied but less capable of greatness and awe. This general vision has expressed itself in specific disquiet about reproductive technologies like cloning and genetic engineering; about the transformation of human embryos into research tools and raw materials; about psychoactive drugs and assorted enhancement technologies; and about a wide array of other attempts to fundamentally reshape human life through biology and medicine. American conservatives have begun to think hard about "where biotechnology may be taking us," as Leon Kass puts it, and what we might do about it.[1]

The resulting intellectual and political activity has melded some of the interests of the pro-life movement with those of conservatives more concerned with the general culture and its institutions, and has formed through that combination an altogether plausible conservative program. This trend, together with several sensational advances in biotechnology, has kept bioethics on the agenda of the American right since the late 1990s. President George W. Bush's first prime-time address to the nation in 2001 was about his new policy on the funding of embryonic stem cell research. Human cloning and related issues have arisen repeatedly in Congress for much of the past decade. And a significant portion of the intellectual energy of the conservative movement has been devoted to the cause of a new bioethics.

And yet, the motives and methods of this movement present conservatives with a profound and complicated problem. Bioethics is necessarily focused on the deepest and most sensitive of human moral intuitions and taboos—those surrounding birth and death, sex and procreation, pleasure and pain, and the meaning of the body. At the same time, it is also directed toward policy, which in a liberal democracy rightly means that it must be an ethics of fully public argument. It is therefore in the business of public argument about taboos—of making the most private things more public, and shining bright lights on things long left in the dark. Herein lies the paradox of a conservative bioethics. Lifting the veil from society's most delicate implicit moral sentiments is hardly a conservative enterprise, and yet one form of doing just that has become a central conservative project. To succeed, a conservative bioethics must be alert to this deep difficulty and its consequences.

DEBATING TABOOS

The word "taboo" was brought into English by Captain James Cook, who heard it used among the Polynesians and marveled at its usefulness. In the original language, taboo describes something that combines in itself both holiness and pollution; it is therefore the most dangerous of all things, and thus forbidden.

Though we in the West have only had the word since 1777, the concept of taboo has always been with us. A taboo is a thing that somehow touches on the venerable, but for that very reason threatens a profound corruption. It stands to profane the highest and most sacred things. It marks a barrier whose violation would strike so deep that we would not have the words to describe it; but we would understand such a violation fully and at once. This unspoken understanding seems always to surround taboos. Speaking of them, bringing them out into the light for all of us to see in detail, is itself seen to put us at grave risk of deep corruption.

Some taboos—like those surrounding incest or cannibalism—are stark and clear, and very nearly universal. The very thought of the corruptions that they represent elicits an almost autonomic revulsion. Others, touching on areas that range from elements of sexuality, to the treatment of the dead and dying, to bodily indignity and even profanity and sacrilege, are of course more controversial. But for those who feel their power, these different taboos all seem to revolve around the avoidance of a deep violation or corruption. What is at stake is not so much the breaking of a rule as the transgressing of a boundary, or a mixing together of things that ought to be kept separate. Taboos stand guard at the border crossings between the realm of the properly human and those of the beasts and the gods. When the boundaries are breached, when degradation or hubris is given expression, our stomachs recoil, even if our minds at first do not.

As Freud points out, an important key to understanding the complicated meaning of taboo is the fact that its opposite in the same Polynesian language is "noa," which means common, or generally accessible.[2] The taboo—part sacred, part unclean—is above all kept out of reach and common view. Its rationale is generally not laid out in detail. We have a sense that deep wisdom is embedded in the prohibition, but that it is better not to unravel it in public. Our most fundamental implicit moral sentiments, which guide us but are themselves best left shrouded, surround and protect our deepest taboos.

These sentiments and insights are reasonable but not fully rational. They are wise but not explicit. We can approach them with arguments but never fully contain them. Try to explain why exactly incest is abominable, and you will find many reasons but probably never quite explain your own abhorrence to complete satisfaction. It is driven by a moral sentiment that you understand but cannot articulate. These sentiments express themselves in almost instinctive responses. For this reason, they are not always reliable, but they are always powerful. And they are also necessary. Though there is often controversy about just where such deeply ingrained limits should be located, it seems clear that no society could function if they were altogether absent. They mark the outer edges of the conscionable, especially with regard to our bodily selves.

Indeed, part of the reason for the inarticulable character of these sentiments is that they very often relate to that element of our existence that is least amenable to rationalization: our embodiment. If we were merely minds, reasoning apart from any body, then our entire experience of life, and the entirety of the ethics that gives us guidance in living well, might be open to fully rational description. But we are embodied creatures, so we can never fully escape, just as we can never fully articulate, the demands made by our bodies on our souls. The sorts of moral insights and taboos that do not lend themselves fully to argument often revolve around parts of our lives to which our embodiment is especially relevant: birth and death; sexuality and pro-creation; bodily wholeness, integrity, and dignity; health and sickness; and family relations, among others. These are the realms where many ethical limits express themselves not in syllogisms but in shudders.

Societies find ways to tiptoe around such taboos. The Greeks told stories about rape and incest, unnatural combinations and inhuman highs and lows, but their stories never simply laid out the matter and explained in full detail the problem with it. They spoke in symbols, hints, and allegories. The Egyptians, the Mesopotamians, and other ancient civilizations had similar tales, and the Biblical religions of course have their own. The image of the sons of Noah walking back-ward (so as not to see) and placing a blanket over their naked father as he sleeps does more than any argument could do to show us the power and hint at the wisdom of our most deeply rooted taboos.[3]

These ancient myths and parables demonstrate a deep awareness of the importance and the danger of taboos, and of the risks of heed-lessly transgressing them or carelessly dragging into full view the implicit understandings that surround them. But modern liberal democracy is notorious for precisely such indiscretion. Or more accu-rately, it prides itself on its ability and willingness to discuss all public questions openly, and lay them out fully for debate before the democratic citizen. Modern democracy may have a greater sense than any of its predecessors of the importance of separating private and public affairs, but everything deemed public (as the questions raised by modern biotechnology have rightly been) is, at least in principle, fully discussed and exposed. For good and bad, very few things are left implicit or unspoken in the life of a liberal democracy.

The greatest teacher of conservatism, Edmund Burke, complained about this tendency of democrats. "It has been the misfortune, not as these gentlemen think it, the glory, of this age, that everything is to be discussed," he wrote.[4] The greatest student of democracy, Alexis de Tocqueville, understood why this should be so, and that often it is for the good. In democratic times, he explains, individuals no longer accept ideas on authority or faith or age-old sentiment. Equality convinces every citizen of the power of his reason, and he wishes to subject every idea to his own rational inspection. Tocqueville describes the public life of a democracy as a constant transformation of the implicit into the explicit, as the authority of tradition and the power of sentiment give way to clearly defined operations of interest and will. Old, deep, unspoken social ties—between owner and tenant, employer and employee, governor and governed, and many others—are transformed into clearly delineated contractual relations, and everywhere old sentimental notions are replaced by explicit arguments. "Do you not perceive on all sides beliefs that give way to reasoning, and sentiments that give way to calculations?" Tocqueville asks.[5]

In our private lives, we democrats surely still respect taboos and may still abide by ancient and unspoken moral intuitions. But in the public life of a democracy, only fully explicated arguments that allow every citizen to consider all the details are finally deemed legitimate. This is just and right and reasonable. But it is also problematic.

It is just and right because we truly cannot and should not depend on moral intuitions and unspoken sentiments to make policy in a democracy. For one thing, these sentiments are unreliable. Repugnance fades with habit. As Dostoevsky warned: man, the beast, gets used to everything.[6] Think of what was deemed unconscionable a generation ago—in art, music, films, public behavior, and the general life of the culture—and then look around. You will now find it everywhere acceptable. For better or for worse, our sentiments can accustom themselves to once unconscionable things, and so they cannot be relied upon alone to guide our conscience.

A second and more important reason not to rely on moral intuitions is that they may simply be wrong or unjust. Interracial marriage, for example, turned the stomachs of many in white America until only very recently. But that gut reaction could not stand up to scrutiny, and should not have been allowed to determine government policy. It is good for us all that it no longer does, but this is so only because arguments (and daily experience) overcame what seemed to many like a deep intuition, and what was indeed a powerful taboo.

A third reason not to rely on our moral intuitions alone is that they do not always draw clear lines for us to follow. Even if we all agreed that a particular taboo or deep repugnance is legitimate and should be heeded, we must still establish a specific policy for doing so, and this still leaves us to argue over the details.

Such democratic argument is good for us. It clarifies important issues, forces all sides to make their best case and engage their opponents, and it is in the end the most just and legitimate way to make public policy. Even as we acknowledge the truth of some of our inarticulate moral sentiments, and even as we live according to them in our private lives, we must also acknowledge that simply codifying them in law would be unacceptable.

And so we argue, and we should. But in some cases, the democratic transformation of sentiments into arguments creates a deep and serious problem. This happens when we must argue in favor of taboos, as a conservative bioethics must often do.

The trouble is not that it is hard to do this. Very often, there are sound and serious arguments to support an old intuition, and these can be marshalled and wielded quite effectively. If something is wrong, it is wrong for a reason, and the reason can be reached by argument. The trouble is that reaching that reason is not itself a neutral process. It has real consequences. It involves unmasking what surrounds the reason, and in the process risks undoing that which the reason defends. The very act of defending taboos in the public arena requires us, in a limited but highly meaningful way, to transgress them—or at least to uncover them in ways that undercut them.

From the beginning, modern conservatives have worried deeply about this difficulty. Edmund Burke believed that the protection of society's hidden intuitions and taboos might be the most urgent task of those concerned for cultural continuity, and their unraveling was among the worst offenses of the French revolutionaries. In an age of such unraveling, Burke wrote,

> All the pleasing illusions, which made power gentle and obedience liberal, which harmonized the different shades of life, and which, by a bland assimilation, incorporated into politics the sentiments which beautify and soften private society, are to be dissolved by this new conquering empire of light and reason. All the decent drapery of life is to be rudely torn off. All the superadded ideas, furnished from the wardrobe of a moral imagination, which the heart owns, and the understanding ratifies, as necessary to cover the defects of our naked, shivering nature, and to raise it to dignity in our own estimation, are to be exploded as ridiculous, absurd, and antiquated fashion.[7]

"What would become of the world," Burke wondered, "if the practice of all moral duties, and the foundations of society, rested upon having their reasons made clear and demonstrative to every individual?"[8] These reasons can be unveiled, to be sure, and they are often good and strong reasons, but they cannot be re-veiled after they have been examined, so a wise society approaches them with great care. Burke insists such care is possible. "Many of our men of speculation," he writes of his fellow Englishmen (in contrast to the French),

> instead of exploding general prejudices, employ their sagacity to discover the latent wisdom which prevails in them. If they find what they seek, and they seldom fail, they think it more wise to continue the prejudice, with the reason involved, than to cast away the coat of prejudice, and to leave nothing but the naked reason; because prejudice, with its reason, has a motive to give action to that reason, and an affection which will give it permanence. Prejudice is of ready application in an emergency; it previously engages the mind in a steady course of wisdom and

virtue, and does not leave the man hesitating in the moment of decision.[9]

But this is mostly wishful thinking, and Burke knew it. The logic of democratic politics demands an unveiling of taboos and an exposing of intuitions, especially when questions of human embodiment and biology are on the table. Indeed, at that dawn of modern conservatism, Burke also understood why the taboos that surround the body, and especially those tied to human natality, were (as we have already seen) those most crucial to defend for any believer in the anthropology of generations. The prerequisites for continuity, and for social cohesion and order, were deeply tied together with the ways in which human generations are connected, Burke believed. The entire conservative conception of society, as Burke presents it, is linked inexorably to this implicit understanding of natality. Burke writes:

> Dark and inscrutable are the ways by which we come into the world. The instincts which give rise to this mysterious process of Nature are not of our making. But out of physical causes, unknown to us, perhaps unknowable, arise moral duties, which, as we are able perfectly to comprehend, we are bound indispensably to perform. Parents may not be consenting to their moral relation; but, consenting or not, they are bound to a long train of burdensome duties towards those with whom they have never made a convention of any sort. Children are not consenting to their relation; but their relation, without their actual consent, binds them to its duties—or rather it implies their consent, because the presumed consent of every rational creature is in unison with the predisposed order of things. Men come in that manner into a community with the social state of their parents, endowed with all the benefits, loaded with all the duties of their situation. If the social ties and ligaments, spun out of those physical relations which are

the elements of the commonwealth, in most cases begin, and always continue, independently of our will, so, without any stipulation on our own part, are we bound by that relation called our country, which comprehends (as it has been well said) "all the charities of all." [10]

In this stirring picture of profound but unchosen human attachments, the inscrutable mysteries at the heart of human natality are what hold society together. And the attempt, therefore, to fully rationalize them, and to sever the generations by exploding in a flash of light the mysterious ties that bind them, is at the heart of the quest for a profound discontinuity in the human experience. "I have read some authors who talk of the generation of mankind as getting rid of an excrement; who lament bitterly their being subject to such a weakness," Burke wrote. "They think they are extremely witty in saying it is a dishonorable action and we are obliged to hide it in the obscurity of night. It is hid it is true: not because it is dishonorable but because it is mysterious." [11]

The dangers of unveiling our deepest taboos in the process of defending them, a danger that has long concerned conservatives, is therefore at its most acute in precisely the areas taken up by modern bioethics. Conservatives cannot avoid this perplexing difficulty, but they will also not find it easy to contend with it. A democratic citizen cannot argue for a taboo as such. *Ignorabimus* makes a bad campaign slogan, and rightly so. The democratic transformation of sentiments into arguments means that not the form or pedigree of the taboo must be defended, but rather the detail of its substance. To undertake such a defense, the substance must be opened up, laid out, and lit up in the glare of the democratic arena.

One consequence of this necessity is the cheapening or profaning of this substance by constant handling and trafficking. Talking about "the moral status of the embryo" the way we talk about tax credits or food safety regulations makes us too familiar with it. By constantly handling it, dealing with it, creating shorthand and acronyms for it, and in general making it a currency of the public debate, we make ourselves less shy, less restrained, and less awed by the deeply meaningful sentiment we are defending. Talk of "pulling the plug," or even "assisted suicide," somehow doesn't leave room for the full human significance of what is involved. Tables comparing the success rates of Gamete Intra-Fallopian Transfer with Intracytoplasmic Sperm Injection blind us to the meaning of the act of artificial procreation. In the fog of bland and banal euphemisms and the flood of bioethics acronyms—IVF, PGD, ICSI, GIFT, ZIFT, SCNT, ESC, ASC, and on and on—moral substance can too easily be obscured.

More importantly, as Burke explained, by transforming a deep moral sentiment into an argument, we abandon and likely lose forever its power as a sentiment. In appealing to a clear and explicit rational argument, we begin to overcome our deep repugnance, and may diminish it in others. We create an argument that rests, as arguments do, upon premises and postulates, rather than a deep taboo drawing on some profound common moral foundation that animates us forcefully but that we cannot adequately put into words. By starting down that road, we point the way toward arguing our taboos out of existence. Each of our premises—even if they are correct—can be undermined by extreme cases or clever manipulations or sheer sophistry, and in the end the subject of our earlier shared repugnance becomes just another controversy, about which differences exist and reasonable people disagree. Some elements of our moral life cannot be completely rationalized, and so appear incomplete in the democratic arena.

In its April 2002 opinion in the case of *Ashcroft v. Free Speech Coalition*, the United States Supreme Court spent pages upon pages trying to articulate an opposition to child pornography, and in the end concluded that it might not be so bad if the pornography is computer-generated or if the actors are only pretending to be children.[12] We begin with an argument in support of a shared conviction, but in the end, by slicing up the sentiment to turn it into palatable arguments, we lose respect for it in its own terms, and the argument that we have crafted becomes just another part of the debate.

HOW MORAL INTUITIONS UNRAVEL

The case of cloning to produce children (or "reproductive cloning") may illustrate this process in action. After the cloning of Dolly the sheep in 1996, the prospect of someday cloning human beings seemed suddenly real. The response was a classic example of a deep moral taboo in action. Over 90 percent of the American people expressed opposition to the notion of cloning children, before any argument, one way or another, had been presented in earnest.[13] Very few people have actually argued since then in favor of cloning children. A few of the usual postmodern sophists offered up a few of the usual postmodern sophistries about perfect freedom and individual will, and here and there some extreme cases were postulated in which cloning might be the only way to do good in a crisis. But even with these few exceptions, deep opposition to reproductive cloning was the rule in the public arena. More or less every member of Congress expressed opposition to it, and polls throughout the late 1990s continued to show that about nine out of ten Americans thought it abhorrent.

Nevertheless, even the staunchest opponents of cloning have not felt comfortable leaving the public consensus at that. In part fearing that it would be undermined by clever arguments, and in part feeling uneasy (and rightfully so) about rooting policy in unarticulated sentiments, cloning opponents set out to articulate explicit arguments against the cloning of children. From issues of safety (which offer only temporary barriers) they moved to explicit discussions of the grounds for the abhorrence of cloning. Arguments about the effect on families, societies, individual identity, and deep cultural norms were presented to make clear and precise the case for refraining from this practice.

While these arguments are effective and powerful, each can be quibbled with on an assortment of grounds. Some depend on views of the family or tradition which are held to be controversial. Some are accused of making unsupported assumptions. Some are said to be unfair or unclear. And the sum of them all still does not describe the full depth of our revulsion at cloning. The practical result of all this is still unknown. Public opinion about cloning children has so far not changed very much. But opponents of cloning have tied themselves to specific concrete arguments which, if they are swept away, can no longer so easily lean on a deep and commonly shared moral taboo. They have transformed a sentiment into an argument, and in the process they may well have begun to undermine the underlying sentiment. The response to assorted (though so far all fake) claims to have cloned a child in recent years showed that some elite opinion-shapers are willing to defend cloning children by rebutting each of the individual arguments against it, leaving aside the more general underlying revulsion.

For good and for bad, this seems to be the fate of moral intuitions in a liberal democracy. The fact that everything must be laid out in the open in the democratic age is destructive of the reverence that gives moral intuition its authority. A deep moral taboo cannot become simply another option among others, which argues its case in the marketplace. Entering the market and laying out its wares takes away from its venerated stature, and its stature is the key to its authority. By the very fact that it becomes open to dispute—its pros and cons tallied up and counted—the taboo slowly ceases to exist. In the long run, this affects not only the public sphere but also our private ethical judgments. The character of democratic culture, which is so good in so many ways, is here slowly corrosive of a vital moral pillar, and one particularly central to the conservative view of the world.

Our country's ablest statesman understood this basic fact of modern politics. In 1838, long before he was president, Abraham Lincoln said this in a speech about the preservation of American institutions:

> Passion has helped us, but can do so no more. It will in future be our enemy. Reason—cold, calculating, unimpassioned reason —must furnish all the materials for our future support and defense. Let those materials be molded into general intelligence, sound morality, and, in particular, a reverence for the Constitution and the laws. . . . Upon these let the proud fabric

of freedom rest, as the rock of its basis; and as truly as has been said of the only greater institution, "the gates of hell shall not prevail against it."[14]

Noble sentiments, he argued, must now be turned into arguments for nobility. Reason must replace passion, but this does not mean that those things about which we are passionate must be lost. They must, instead, be defended rationally and explicitly, and they can be.

Twenty-three years later, however, Lincoln seemed less certain of this. In another speech about preserving our institutions, this one his first inaugural address in 1861, he again made the case for reverence for the laws, sound morality, and freedom, but he ended with the following appeal:

> Though passion may have strained, it must not break our bonds of affection. The mystic chords of memory, stretching from every battlefield and patriot grave to every living heart and hearthstone all over the broad land, will yet swell the chorus of the Union when again touched, as surely they will be, by the better angels of our nature.[14]

Mystic memory is very different from cold reason, and in our moment of greatest crisis, this greatest democratic statesman sensed that arguments alone were not enough to hold the bonds of a society together. Strong as they are, and they can be strong indeed, arguments are finally not a fully satisfactory substitute for moral intuitions and untouchable sentiments.

In our time, on most issues and especially in the science debates, conservatives neither have nor desire to have recourse to such mysticism. And so they must argue, knowing they are doing away with the foundations of the old, and as they do so struggling to construct new foundations—shallower but hopefully firm—before the structure topples over. They are active participants in the process of diminishing the influence of moral intuitions and replacing them with what they hope are strong arguments, but what deep down they fear are only temporary barriers against the nihilistic force of cynical relativism. They are engaged in a most unconservative project: dissecting taboos, hoping to save something of them.

FROM BIOLOGY TO ETHICS

There is no avoiding this precarious project, and the language of dissection begins to show us why, by pointing us to something interesting and important about the biotechnology debates. It shows us that the taking apart of taboos, and the dragging of the hidden into the open, is not only a challenge that confronts us as we argue, but also what has drawn us into the argument to begin with. After all, this process of making the implicit explicit, and then open to question and manipulation, is what modern biology itself does.

Human biology was also once a realm of semi-mystery, whose causes and components remained mostly hidden. We studied the body and sought to treat its ailments, but it remained a coherent whole, whose most primal workings were not open to our inspection.

Modern biology is different. It works, and works well and to our benefit, by studying living things in the laboratory, outside their natural contexts, by taking them apart and examining components down to the genetic and molecular level. Modern biology allows us to open ourselves up, to make the mysterious known in detail, and to tinker and manipulate. Modern science, like modern politics, functions by bringing everything into the light.

This approach, and its significance, is most apparent with regard to what might once have been the most mysterious realm of human biology and what is now the most controversial arena of bioethics: procreation and human origins.

Embryologists in the laboratory are, quite literally, dissecting taboos. The embryo is the perfect physical example of the taboo: undifferentiated holiness and pollution, at once both awesome and profoundly dangerous. To see it in full view is to sense that it was never meant to be looked upon. The strange old Jewish description of embryonic life, which compared it to liquid, now strikes us as very odd and maybe even insufficiently respectful. But it aptly captures the sense that this developing life once seemed to us amorphous, mysterious, non-specific, and above all unknown. This approach to the embryo—though proven false by modern embryology—did allow us to afford it a genuine, deep, and implicit respect.

But all that has changed. As famed biochemist James Watson told a congressional committee in 1971: "Human embryological development need no longer be a process shrouded in secrecy. It can become instead an event wide open to a variety of experimental manipulations."[16] The shroud is gone, and today we know the embryo in great detail. Indeed, it may be that we know it too well to respect it. We either disrespect it, or we must construct an intricate rational argument, based on precisely our intimate knowledge of its biology, that claims an utmost confidence about what the embryo must be to us, or why it is "one of us." Such an argument does call for protection, but it cannot call for fully human reverence.

The effect of both approaches is at its most acute (if not absurd) when the cause—our explicit knowledge of the embryo—is most developed: in the case of the embryo that exists in the laboratory, outside the body of its mother.

The existence of this strange being called the extra-corporeal embryo is what forces us into many of the strangest, most heated, and most profound of the science debates. The extra-corporeal embryo has been ripped from its human context, in which the very early embryo is not quite a distinct player in itself but a deeply embedded and mostly unknown potential. It has been put before us to be considered in isolation, where it barely makes any sense. We look at this creature, which has been manufactured, molded, formed, examined, and up to a certain point developed under the lights of the laboratory. It is growing, but can only grow so far without further biotechnical intervention. It is living, but only because the scientists have created it artificially. It is human, though our eyes may deny it. It is useful as a resource for medical research, but is no less a human being than it ever will be. What in the world are we supposed to do with this thing? How is ethics supposed to serve us in this circumstance?

The moral challenge of this situation is so vexing because its central problems do not arise in any inherent way out of normal human experience, and therefore are not well served by a moral philosophy built around that experience. They confront us because we have made the implicit, mysterious, original form of the human creature into an explicit, carefully studied, painstakingly examined object of scientific inquiry. We know so much about it already that our usual ways of dealing with it—ways that revolve around an implicit respect—can no longer be adequate.

We react to it with an attempt to practice sound ethical reasoning. We ask questions: How shall we regard this thing? Is it one of us? Does it have moral standing? Does it have rights, or shall we use it for our own ends? What strange questions to ask about such a thing! And yet they are the right questions, and we are right to ask them. They are absurd only because the situation is absurd, and it is so precisely because we have turned an implicitly mysterious taboo into an explicitly known and meticulously scrutinized object.

To answer these questions, we need to get to know the embryo even better than we already do. We need to understand its development more clearly; we need to comprehend its potential viability, its genetic characteristics, its physical form. There is no turning back once we have given up on the taboo. And the embryo debate without taboos begins to overflow with outlandish ironies. The embryo's most adamant defenders argue in favor of the inviolability of human embryos by resorting to the latest detailed scientific data and analysis, some of it obtained by taking embryos apart. Meanwhile those desperately seeking to use it as a resource for research argue that the embryo is not worth much. The Catholic bishops release statements making reference to arcane articles in scientific journals while the scientists tell us that the embryo is no larger than the period that ends this sentence, so we should not trouble ourselves over it.

Though the bishops' mode of argument is clearly more responsible, neither is finally satisfactory. Both sides argue over what the science tells us we should think about the embryo, and neither now can speak to an implicit moral intuition. There is no clearer case of a profaning of the sacred in our time, and no clearer example of the consequences of dissecting taboos. And yet given how far we have come, there is no choice but to proceed this way. The issues at stake are too important to ignore, and so conservatives cannot simply abandon the debate to avoid dismembering moral intuitions.

CONSERVATIVE FUTURISM

A public bioethics is therefore unquestionably necessary. Conservatives have no choice but to participate in its development, but in doing so they must keep in mind the stakes, and remember that this project holds in its hands the future not only of public thinking about the questions opened up by the new biology, but in the long run also private thinking on the subject. Even in the best of cases—where its task is to enshrine in policy the substance of nearly universal moral sentiments—it must proceed by undoing these sentiments and replacing them with democratic substitutes that may be more effective in public life but could never be as strong.

This is why the task confronting the right in the science debates is sometimes so difficult. It must transform moral sentiments into arguments for morality. Its chief ally in this effort is the deep moral wisdom at the heart of our civilization—by which most Americans live their lives. But the effort itself can pose real risks to precisely the character of that wisdom.

The nature of both modern science and modern politics demands that the argument proceed this way. Both incessantly unveil the veiled and shine light on hidden things. We gain much that is immensely beneficial from both, but we risk losing much if the process of transforming sentiments into arguments is not carried out properly, in a sober and responsible way, and with an eye to what is worth preserving and protecting. Bioethics, at the juncture of politics and science, is where the struggle for the character of the new biotechnological age will be waged. And conservatives are right to enter the fray as they have.

Having entered, however, they will not find it easy to win. Conservatism traditionally leans on and seeks to protect the implicit wisdom contained in age-old institutions and social arrangements. It goes beyond this of course, and makes arguments and is at home in liberal democratic politics. But much of its appeal, and many of its arguments, are rooted in a sense that certain of the old assumptions have some value and some truth. In these science debates, however, the right is sometimes forced to proceed by pulling up its own roots, and to begin by violating some of the very principles it seeks to defend. To do this without self-destructing, it must understand that in the long run this responsible replacement of sentiment by argument is a key component of its mission. It must therefore begin to lean less on implicit intuitions and develop for itself a very clear and explicit sense of the world it seeks to defend, and the dangers it seeks to avert.

What results from these reflections is therefore not an argument against argument, but rather quite the contrary, a call for more and better argument. In the long run, a conservative bioethics must lay out a more fully developed positive vision of what is worth defending and why, and a more thoroughly articulated negative vision of the dangers that confront us if the "Brave New World" becomes reality. Making this case requires arguments from first principles, and not just reactions to individual technologies or fearful insinuations. Code words like "designer babies" will only have an impact as long as there are still deep-seated taboos and implicit intuitions to which they can appeal. With time, these intuitions may weaken dramatically—perhaps they already have—and conservatives must be ready with another arsenal of arguments.

The present task of the right in the science debates (and especially the biotechnology debates) is therefore to develop and articulate a coherent worldview—to put meat on the bones of loosely defined terms like "human dignity" and "Brave New World" and turn ethical disquiet into public arguments. It must explore the character of the changes made likely by science and technology, with an eye to their effect on our attitudes about ourselves, our dispositions toward our bodies and souls, our sense of the appropriate uses and limits of human power, the shape of the future, the prospects for cultural continuity, and the form and function of our society. It must ask what sort of world we are creating for ourselves, and what sort of place it will be for future generations to enter and inhabit. It should begin from a sense of what is humanly important, and try to envision, in a rigorously informed but imaginative way, the path laid before us by the logic driving scientific developments. Knowing that precise prediction is pure folly, but that informed forethought is essential, it must construct for itself an approximate sense of what the future may plausibly bring, and which among the possibilities should be avoided or encouraged and why. This means that it must engage in some hard thinking in the years to come.

But as it does so, it must also engage in some hard politics. The very process of defending their ideals will make the task of conservatives in the biotechnology debates increasingly difficult, but those who carry out this task should note that implicit sentiments and intuitions are being sucked out of our public life much faster than they are disappearing from our private lives. While taboos may have less force in public argument and policymaking, the arguments crafted to replace them will still appeal to many people whose own souls have not lost the ability to feel an inarticulable awe. They, too, understand the absolute (and quite reasonable) need for arguments. But they will find the conservative argument especially attractive, if it is properly formulated and expressed. This means that in the short term, as it works to formulate its worldview, a conservative bioethics should also seek to reinforce the cultural institutions that support vital common sentiments, to use the force of public opinion to make practical inroads, and to be engaged in politics, not just contemplation.

Indeed, politics and contemplation are closely intertwined in the service of the same general ends. The pressures and urgencies of political debate can force hard thinking that might otherwise have been avoided. When the stakes are clear, and the debate comes down to a vote, the need to make arguments is at its most stark. This can sometimes hurt the cause of clarity and understanding, by forcing the partisans to make dishonest or underhanded appeals, but it can also help that cause by forcing serious people to think through the issues involved in a serious way and to make their best case. The bioethics debates of the past decade have resulted in more and better writing about the fundamental issues underlying bioethics than had been seen in at least the previous decade, and especially on the right. A concrete political choice imbues the debate with a sense of responsibility and with sharp focus.

If this combination of intellectual, cultural, and political work is carried off well, the right has a chance to make its concerns felt in the science debates. But carrying it off well will require a sense of the risks and the exigencies that result from the self-immolating character of the project. Uprooting moral intuitions in the cause of moral living will not be easy, and the dangers for the right along the way are great. For the right, no less than the left, the science debates force to the surface some profound contradictions.

Conclusion

PROPERLY UNDERSTOOD, America's science debates are about the most essential question of our politics: the question of the future. It is the nature of the modern scientific enterprise to put this question most starkly and forcefully before our democratic system, and therefore to force to the surface some crucial facts and problems. The divide that gives our political life its basic shape turns out, when examined in light of the science debates, to be a disagreement about whether the most important thing about the future is the pursuit of material innovation or the safeguarding of cultural continuity. The left and the right each has its answer, and each answer turns out to be deeply problematic in its own way.

The left takes science to be "on its side"—and with some reason— but science is also profoundly dangerous to the left's own causes and core beliefs. In its aggressive defense of the prerogatives of science, therefore, the left risks losing sight of the difference between material and moral advancement, and so undercutting the very ideals it seeks to champion. The right meanwhile sees science as essential but dangerous, and so as in need of moderation. But the task of moderating it risks both wounding the scientific enterprise and eviscerating the right's own highest aims and purposes.

All of this highlights some fierce tensions in our political life, and especially within the worldviews of the right and the left. These tensions are made most apparent in light of the science debates, but they are not fundamentally about science, but about the political life of our democratic republic.

A book like this one runs the risk of suggesting that science and technology are the lenses through which all things are to be understood, or the essence of modern life. This kind of scientific essentialism, arguing that modern technology makes the present altogether different from the past and so renders the questions of the past irrelevant and obsolete, has been a common strand of thinking in our time, and it has led to many foolish errors and dead ends, and above all to the attempt to eradicate political ideas from the study of political life. There is perhaps no better example of this kind of excessive determinism than German philosopher Martin Heidegger's contention, in 1949, that "agriculture is now a mechanized food industry, so as for its essence it is the same thing as the manufacture of corpses in the gas chambers and the death camps."[1] The technological character of the Nazi horror was the essence of its gruesome inhumanity, Heidegger suggests, and so any practice that shares in that character must share also in that inhumanity. This is surely among the most foolish and ridiculous assertions in the history of Western philosophy, but it gestures toward a thread of political thinking that is very much alive in certain corners of both the left and the right—where technological determinism, disappointment with modern life, and a fervent quest for an elusive authenticity combine to yield a terrible confusion.

What Heidegger saw as the essence of things is not the essence at all. The essence, as ever, is human flourishing—the potential for it, the quest to define and redefine and achieve it, the desire for it, the obstacles to it. It is by that criterion that we should judge the meaning of science and technology in our society (and of course by that criterion we can easily judge the Nazi horrors altogether differently from mechanized agriculture). Human flourishing is the rich and very complicated end toward which our various means must reach. It is what our politics at its best exists to serve—and our science, too. Their interaction in that service has been the subject of this book. And the lesson we are left with is that the dangers science poses to our politics could be best addressed by understanding science properly as a great human endeavor: offering vast promise, but appropriately answerable to the larger society, and legitimately subject to its democratically established rules and boundaries. Science will not overwhelm our faith in equality and our commitment to the essential prerequisites for progress unless we lose our confidence in the authority of our institutions of self-government, and in the truths that underlie them.

The science debates therefore force us to confront some serious challenges to American self-government. To meet these challenges, America must moderate its hopes and fears alike, and reinvigorate its faith in its political traditions. It must not lose sight of the careful balance it has always sought between material advancement and moral progress; the protection of our own rights and liberties, and the passing down to our descendants of the great traditions we inherited. This is neither quite a conservative nor a liberal project. It is the American project, and it is grounded in the view that concern for present freedom and regard for future generations are not two aims but one. In the science debates, no less than in our common life in general, we would do well to remember that unity of aims, and to recall the great and difficult charge that still echoes in our ears in the steady hopeful voices of our nation's founders: to secure the blessings of liberty to ourselves and our posterity.

ACKNOWLEDGMENTS

This book owes its life, from beginning to end, to *The New Atlantis* magazine. Most of its chapters first appeared in some form in the magazine's pages, and all of its arguments were developed or honed with the help of the magazine's editors. To be grateful to *The New Atlantis* is to be grateful to Eric Cohen and to Adam Keiper. Eric, who founded the magazine and imbued it with his special spirit, is a man of vision and imagination, moral clarity, good humor, and class. And more importantly, he is a treasured friend. Adam, who managed the magazine from its inception and now is its editor as well, is a completely implausible mix of raw intelligence, energy, curiosity, conviviality, intellectual depth, and good sense. He is also my oldest and dearest friend, and I do not believe I have had a single useful insight or idea since junior high school that was not either conceived or vastly improved in conversation with Adam. The magazine would also be adrift without its able assistant editor Caitrin Nicol, to whom I am also very grateful for assistance with this book.

The Ethics and Public Policy Center in Washington provides a home not only to *The New Atlantis* but also to me. It is an island of collegiality and deep commitment to American ideals in a city where both are too often lacking, and for that I am immensely grateful to its president, Ed Whelan, and to my colleagues.

A number of other advisers and friends contributed valuable insights, suggestions, and corrections in the course of this project. They include Carter Snead, Gilbert Meilaender, Robert George, William Hurlbut, Paul McHugh, Adam Wolfson, Andrew Bremberg, and Christine Rosen.

But easily the greatest debt I have incurred in this effort (as in so many others) is to Leon Kass. He is, to begin with, my teacher, and if this book says anything at all of any value it is because I learned it from him. But more than that, he is a model of intellectual honesty and rigor, of moral fiber and character, of devotion to family and country, of reasoned and responsible public engagement, and of sheer generosity of spirit. I have been blessed to know him.

This book would not have been possible without the help of all these individuals, though its many imperfections are of course my fault alone. But there is more to life than books, and much more to be grateful for than a project completed. My greatest fortune has been my family. My parents, Gila and Gad Levin, and my brother and sister are owed more than I could offer here. And my wife, Cecelia, is a kind of constant proof for me that we human beings sometimes get more in life than we deserve. She is a miracle. I am grateful to her, and for her, above all; and I dedicate this book to her with love, and with a smile.

NOTES

CHAPTER ONE · THE MORAL CHALLENGE OF
MODERN SCIENCE

1 President George W. Bush, Address, "Remarks at the Medical College of Wisconsin in Milwaukee, Wisconsin, February 11, 2002," in *Public Papers of the Presidents of the United States* (Washington, D.C.: United States Government Printing Office, 2002), XLIII: 232.

2 Albert Einstein, "Science and Religion," in Bronstein, *Approaches to the Philosophy of Religion: A Book of Readings* (New York: Ayers Publishing, 1954), 68–69.

3 Francis Bacon, *The Wisdom of the Ancients* (Whitefish, MT: Kessinger Publishing, 1992), 246.

4 Francis Bacon, *Francis Bacon: A Selection of His Works*, ed. Sidney Warhaft (London: Macmillan & Co., 1965), 374.

5 Francis Bacon, *The Advancement of Learning* (Chicago: Encyclopedia Britannica Inc., 1952), 16.

6 René Descartes, *Discourse on Method*, trans. Richard Kennington (Newburyport, MA: Focus Publishing, 2007), 49.

7 National Academy of Sciences, *Scientific and Medical Aspects of Human Reproductive Cloning.* (Washington, D.C.: National Academies Press, 2002), 1.

8 Descartes, *Discourse on Method*, 49.

9 Aristotle, *The Politics*, I.2, 1252b27–1253a18.

10 Niccolò Machiavelli, *Discourses on Livy*, trans. Harvey C. Mansfield and Nathan Tarcov (Chicago: University of Chicago Press, 1996), 7.

11 House Committee on Government Reform, Subcommittee on Criminal Justice, Drug Policy, and Human Resources, *RU-486: Demonstrating a Low Standard for Women's Health?*, 109th Cong., 2nd Session, May 17, 2006.

12 Sheryl Gay Stolberg, "Method Equalizes Stem Cell Debate," *New York Times*, November 21, 2007.

13 Barack Obama, "Obama Statement on Bush Veto of Stem Cell Bill," Statement released by the Office of Senator Barack Obama, June 20, 2007.

14 President John F. Kennedy, Address, "Remarks to Members of the White House Conference on National Economic Issues – May 21, 1962" in *Public Papers of the Presidents of the United States* (Washington, D.C.: United States Government Printing Office, 1962), XXXV: 203.

15 Strickland was speaking at a June 19, 2001 hearing of the Energy and Commerce Committee of the U.S. House of Representatives. See Francis Fukuyama, *Our Posthuman Future* (New York: Picador, 2003), 185.

16 Marie Jean Antoine Nicolas de Caritat, Marquis de Condorcet, *Sketch for a Historical Picture of the Progress of the Mind* (New York: Noonday Press, 1955), 173.

17 Auguste Comte, *Course in Positive Philosophy* (New York: Barnes & Noble, 1974), 138.

18 Jean-Jacques Rousseau, *The First and Second Discourses*, trans. R. Masters and J. Masters (New York: St. Martin's Press, 1969), 56.

19 Michel Montaigne, *Essays*, trans. C. Cotton (New York: Viking Press, 1993), 77.

20 Genesis 11:4.

CHAPTER TWO · THE CRISIS OF EVERYDAY LIFE

1 Dana Milbank, "Of Stem Cells and Heartstrings," *Washington Post*, April 12, 2007, A2.

2 "Testimony of James Cordy on behalf of the Coalition for the Advancement of Medical Research before the Senate Appropriations Committee, Subcommittee on Labor, Health and Human Services, and Education; May 22, 2003." Made available by the Coalition for the Advancement of Medical Research.

3 Kristen Philipkoski, "Senator Pushes Adult Stem Cells," *Wired*, July 15, 2004, http://www.wired.com/medtech/health/news/2004/07/64221.

4 Bacon, *Francis Bacon: A Selection of His Works*, 320.

5 Michael Sandel, *The Case Against Perfection* (Cambridge, MA: Harvard University Press, 2007), 119.

6 Harvey Mansfield, "Burke's Conservatism," in *An Imaginative Whig*, ed. Ian Crowe (Columbia, MO: University of Missouri Press, 2005), 63.

7 Thomas Hobbes, *Leviathan* (Cambridge: Cambridge University Press, 2000), 70.

8 Alexis de Tocqueville, *Democracy in America*, trans. Harvey C. Mansfield and Delba Winthrop (Chicago: University of Chicago Press, 2000), 513.

CHAPTER THREE · TWO CULTURES?

1 C. P. Snow, *The Two Cultures* (Cambridge: Cambridge University Press, 1993), 9.

2 *Ibid.*, 11.

3 *Ibid.*, 4.

4 *Ibid.*, 5.

5 *Ibid.*, 22.

6 *Ibid.*, 11.

7 *Ibid.*, 25.

8 *Ibid.*, 48.

9 *Ibid.*, 11.

10 C. P. Snow, "The Two Cultures: A Second Look," reprinted in C. P. Snow, *The Two Cultures* (Cambridge: Cambridge University Press, 1993), 79.

11 F. R. Leavis, *Two Cultures: The Significance of C. P. Snow* (New York: Pantheon Books, 1963), 11.

12 *Ibid.*, 29.

13 Lionel Trilling, "Science, Literature & Culture: A Comment on the Leavis-Snow Controversy," *Commentary* 33, no. 6 (June 1962), 462.

14 *Ibid.*, 470.

15 *Ibid.*, 473.

16 *Ibid.*

17 Snow, *The Two Cultures*, 101n6.

18 William Galston, "Liberal Democracy and the Problem of Technology," in *Technology in the Western Political Tradition*, ed. Arthur Melzer, Jerry Weinberger, and Richard Zinman (Ithaca, NY: Cornell University Press, 1993), 249.

19 See, for instance, Steven Pinker, "The Stupidity of Dignity," *The New Republic*, May 28, 2008.

CHAPTER FOUR · TWO VISIONS OF THE FUTURE

1 Virginia Postrel, *The Future and Its Enemies* (New York: Touchstone, 1999), 37.

2 Frederick Hayek, *The Constitution of Liberty* (London: Routledge, 1960), 29.

3 Hans Jonas, The Imperative of Responsibility (Chicago: University of Chicago Press, 1985).

4 Churchill made the remark in a speech on January 3, 1920, noted in Markku Ruotsila, *British and American Anti-communism Before the Cold War* (London: Routledge, 2001), 170.

5 Mihail Roco and William Sims Bainbridge, eds., *Converging Technologies for Improving Human Performance* (Washington, D.C.: National Science Foundation, 2002), 6.

6 Max More, "On Becoming Posthuman," http://www.maxmore.com/becoming.htm.

7 William Godwin, "Enquiry Concerning Political Justice," reprinted in *The Economics of Population: Classic Writings*, ed. Julian Simon (London: Transaction Publishers, 1998), 38.

8 *Ibid.*

9 See especially Hannah Arendt, *The Human Condition* (Chicago: University of Chicago Press, 1958).

10 Edmund Burke, *The Writings and Speeches of Edmund Burke*, ed. Paul Langford (Oxford: Oxford University Press, 1988), VIII: 147.

11 Thomas Jefferson, "Letter to James Madison, September 6, 1789" in *Writings* (New York: Library of America, 1984), 959.

12 "Implications of Cloning," *Nature* 380 (1996), 383.

13 Hannah Arendt, *Between Past and Future* (New York: Penguin, 1993), 192-193.

14 See Plato, *The Republic*, books V and VI.

15 Ronald Bailey, "Is Freedom Just Another Word for Random Genes?," http://www.reason.com/news/show/34791.html.

16 C. S. Lewis, *The Abolition of Man* (New York: Harper Collins, 2001), 56-57.

17 Hans Jonas, *The Imperative of Responsibility* (Chicago: University of Chicago Press, 1985), x.

CHAPTER FIVE · SCIENCE AND THE LEFT

1 Hillary Clinton, "Scientific Integrity and Innovation: Remarks at the Carnegie Institution for Science," speech transcript made available by the office of Senator Hillary Clinton.

2 Ruth Walker, "Kennedy praises MIT, decries White House pseudo-science," *MIT Tech Talk*, April 25, 2007, 3.

3 Emily Esterson, "Richardson Clambakes in New Hampshire," Associated Press, May 7, 2007.

4 Thomas Paine, *The Complete Writings of Thomas Paine*, ed. Phillip Foner (New York: Citadel Press, 1945), II, 242.

5 John Dewey, *The Later Works: 1925-1953* (Carbondale, IL: Southern Illinois University Press, 1984), vol. 2, 379.

6 Descartes, *Discourse on Method*, 49.

7 Bacon, *Francis Bacon: A Selection of His Works*, 320.

8 *Ibid.*, 331.

9 John Muir, *My First Summer in the Sierra* (Boston: Houghton Mifflin, 1911), 110.

10 Barbara Ward, *Only One Earth* (London: Andre Deutsch, 1972), 17.

11 Alan Weisman, *The World Without Us* (New York: Thomas Dunne Books, 2007), 4.

12 Charles T. Rubin, *The Green Crusade* (New York: Rowman & Littlefield, 1994), 10.

13 Dan Balz and Juliet Eilperin, "Gore and U.N. Panel Share Peace Prize," *Washington Post*, October 13, 2007, A1.

14 James Bone, "Britain Puts Climate Change on UN Agenda," *Times of London*, March 8, 2007, 3.

15 "European President Says Climate Change EU's Top Priority," Environmental News Service, May 25, 2007, http://www.ens-newswire.com/ens/may2007/2007-05-25-03.asp.

16 Erik Kirschbaum, "Merkel rejects call to moderate emissions cuts," Reuters, July 3, 2007.

17 Thomas Jefferson, *The Works of Thomas Jefferson*, ed. Paul L. Ford (New York: Putnam & Sons, 1905), 477.

18 John Locke, *Second Treatise of Government* (Indianapolis, IN: Hackett Publishing, 1980), 8.

19 David Hume, *An Enquiry Concerning the Principles of Morals* (Indianapolis, IN: Hackett Publishing, 1983), 25.

20 Herbert Croly, *The Promise of American Life* (New York: Macmillan, 1909), 400.

21 Margaret Sanger, *The Selected Papers of Margaret Sanger*, ed. Esther Katz (Champaign, IL: University of Illinois Press, 2003), 321.

22 C. Mansfield *et al.*, "Termination rates after prenatal diagnosis of Down syndrome, spina bifida, anencephaly, and Turner and Klinefelter syndromes: a systematic literature review," *Prenatal Diagnosis*, September 1999, 808–812.

23 E. Garne *et al.*, "Prenatal diagnosis of severe structural congenital malformations in Europe," *Ultrasound Obstetrics and Gynecology*, January 2005, 6–11.

24 Lois Rogers, "Having Disabled Babies Will Be 'Sin,' Says Scientist," *Sunday Times*, July 4, 1999.

25 Johann Hari, "Why I Support Liberal Eugenics," *The Independent*, July 6, 2006.

26 Allen Buchanan, Dan W. Brock, Norman Daniels, and Daniel Wikler, *Chance to Choice* (Cambridge: Cambridge University Press, 2000), 43.

27 John Rawls, *A Theory of Justice* (Cambridge, MA: Harvard University Press, 2005), 108.

28 Some of the most extreme opponents of equality on the American scene have made this point too. Princeton "ethicist" Peter Singer (known for his advocacy of euthanizing disabled newborns) noted in a letter to the *New York Times* (June 20, 2005) that "The crucial moral question is not when human life begins, but when human life reaches the point at which it merits protection. . . . Unless we separate these two questions—when does life begin, and when does it merit protection?—we are unlikely to achieve any clarity about the moral status of embryos."

CHAPTER SIX · SCIENCE AND THE RIGHT

1 Leon Kass, "Ageless Bodies, Happy Souls," *The New Atlantis*, no. 1 (Spring 2003), 9.

2 Sigmund Freud, *Totem and Taboo*, trans. James Strachey (New York: W.W. Norton, 1950), 19.

3 Genesis 9:20–23. Every reader with an interest in this perplexing subject is strongly encouraged to read Leon Kass's insightful discussion of Noah and his sons in *The Beginning of Wisdom: Reading Genesis*, (New York: Free Press, 2003), Chapter 7.

4 Burke, *The Writings and Speeches of Edmund Burke*, 142.

5 Tocqueville, *Democracy in America*, 228.

6 Fyodor Dostoevsky, *Crime and Punishment*, trans. Constance Garnett (New York: Collector's Library, 2004), 47.

7 Burke, *The Writings and Speeches of Edmund Burke*, 128.

8 *Ibid.*, I: 136.

9 *Ibid.*, VIII: 138.

10 Edmund Burke, *Selected Writings and Speeches of Edmund Burke*, ed. Peter Stanlis (London: Transaction Publishers, 2006), 648.

11 Edmund Burke, *A Note-Book of Edmund Burke*, ed. H. F. V. Somerset (Cambridge: Cambridge University Press, 1957), 91-92.

12 *Ashcroft v. Free Speech Coalition*, 535 U.S. 234 (2002). For a similar example, see H.R. 3687, proposed in the 108th Congress (2003) to establish restrictions on profanity in broadcasting, and available through http://thomas.loc.gov.

13 Gail Javitt *et al.*, "Cloning: A Policy Analysis," Genetics and Public Policy Center, 2005.

14 Abraham Lincoln, *Speeches and Writings*, ed. Roy Basler (New York: De Capo Press, 2001), 84.

15 *Ibid.*, 588.

16 James Watson, *A Passion for DNA* (Oxford: Oxford University Press, 2000), 84.

CONCLUSION

1 Heidegger made the remark in a lecture entitled "The Frame" in 1949 (see Victor Frarias, *Heidegger and Nazism* [Philadelphia: Temple University Press, 1989], 287). The lecture was later turned into a portion of Heidegger's book *The Question Concerning Technology*, but this particular remark was removed.

INDEX

A NOTE ON THE TYPE

IMAGINING THE FUTURE has been set in Minion, a type designed by Robert Slimbach in 1990. An offshoot of the designer's researches during the development of Adobe Garamond, Minion hybridized the characteristics of numerous Renaissance sources into a single calligraphic hand. Unlike many early faces developed exclusively for digital typesetting, drawings for Minion were transferred to the computer early in the design phase, preserving much of the freshness of the original concept. Conceived with an eye toward overall harmony, Minion's capitals, lowercase letters, and numerals were carefully balanced to maintain a well-groomed "family" appearance—both between roman and italic and across the full range of weights. A decidedly contemporary face, Minion makes free use of the qualities Slimbach found most appealing in the types of the fifteenth and sixteenth centuries. Crisp drawing and a narrow set width make Minion an economical and easygoing book type, and even its name evokes its adaptable, affable, and almost self-effacing nature, referring as it does to a small size of type, a faithful or favored servant, and a kind of peach.

SERIES DESIGN BY CARL W. SCARBROUGH